EVOLUTIONARY PROTISTOLOGY

The Organism as Cell

Proceedings of the 5th Meeting of the International Society for Evolutionary Protistology, Banyuls-sur-Mer, France, June 1983

Edited by

LYNN MARGULIS
Boston University

MARIE-ODILE SOYER-GOBILLARD
Laboratoire Arago

JOHN CORLISS
University of Maryland

Reprinted from *Origins of Life*, Volume 13, Nos. 3-4

D. Reidel Publishing Company

Dordrecht / Boston

ISBN-13: 978-94-009-6400-6   e-ISBN-13: 978-94-009-6398-6
DOI: 10.1007/978-94-009-6398-6

Softcover reprint of the hardcover 1st edition 1984

TABLE OF CONTENTS

1. Michel Herzog
2. Paul Prevot
3. Colette Galleron
4. F. J. R. Taylor
5. Marie-Odile Soyer
6. Annie Boillot
7. Yvonne Bhaud
8. Peter Rizzo
9. Peter Kivic
10. David Sigee
11. Anton Bopman
12. John Dodge
13. Christine Metivier
14. Marie Albert
15. Guy Brugerolle
16. Pilar Garcia
17. Isabella Desportes
18. M. Dynner
19. Martin Estrada
20. Øjvind Moestrup
21. J.-P. Mignot
22. Marcella Lefotron
23. David Patterson
24. Claude Grevet
25. Peter Westbroek
26. Urs Peter Roos
27. Colette Demar
28. Veronique Machelon
29. Patricia Walne
30. John Corliss
31. Alvin Poppenheimer
32. Andre Lwoff
33. Alain Bilbaut
34. Michelle Laval-Peuto
35. Jean Genermont
36. Deniis Searcy
37. Guy Orrison
38. Josette Rouvière-Yaniv
39. Jean Cachon
40. I. Brent Heath
41. Betsy Dexter-Dyer
42. Robert Allen
43. Gisele Bardele
44. Mrs. Motta
45. M. C. Tellez
46. Monique Cachon
47. Thos Cavalier-Smith
48. Richard Ludueña
49. Faye Murrin
50. Günther Bahnweg
51. Michael Melkonian
52. S. T. Moss
53. John Heisler
54. Andre Adoutte
55. Gil Rosano
56. Lisa MacKerracher
57. Susan Lenk
58. Morris Alexander
59. Melvyn Little
60. Diana Lipscomb
61. David Lloyd
62. Lynn Margulis
63. Jerome Motta
64. Dennis Goode
65. Christian Bandele
66. Vincent Demoulin
67. Eugene Small
68. Jean Febvre

LIST OF PARTICIPANTS

I.S.E.P. 1983

Adoutte André, Centre de génétique moléculaire, C.N.R.S., 91190 Gif-sur-Yvette (France)
Albert Marie, Laboratoire Arago, 66650 Banyuls-sur-Mer (France)
Allen Robert D., Department of Biological Science, Dartmouth College, Hanover, NH, 03755 (U.S.A.)
Bahnweg Gunther, Institut für Mikrobiologie der Universität Göttingen, Grisebachstrasse 8, 34 Göttingen (F.R.G.)
Bardele C. F., Institut für Biologie III, Lehrstuhl Zoologie, Auf der Morgenstelle 74, Tübingen (F.R.G.)
Bhaud Yvonne, Laboratoire Arago, 66650 Banyuls-sur-Mer (France)
Bilbaut A., Laboratoire d'Histologie et Biologie tissulaire, Université Claude Bernard - Lyon I, 43 Boulevard du 11 novembre, 69621 Villeurbanne (France)
Billy (de) Françoise, Laboratoire Arago, 66650 Banyuls-sur-Mer (France)
Boillot Annie, Laboratoire Zoologique, La Darse, 06300 Villefranche-sur-Mer (France)
Boletzky Sigurd (von), Laboratoire Arago, 66650 Banyuls-sur-Mer (France)
Borman Anton H., Biochemisch Laboratorium, Rijksuniversiteit Leiden, Wassenaarseweg 64 2333 AL Leiden (Holland)
Brugerolle Guy, Laboratoire de Zoologie, Complexe scientifique des Cézeaux, B.P. 45, Les Cézeaux, 63170 Aubière (France)
Cachon Jean, Laboratoire de Protistologie marine, Station de Biologie marine, 06230 Villefranche-sur-Mer (France)
Cachon Monique, Laboratoire de Protistologie marine, Station de Biologie marine, 06230 Villefranche-sur-Mer (France)
Cavalier-Smith T., University of London Kings College, 26 Drury Lane, London WC2B 5RL (U.K.)
Corliss John O., Department of Zoology, University of Maryland, College Park, MD 20742 (U.S.A.)
Demar Colette, Laboratoire de Zoologie, 46 rue d'Ulm, 75230 Paris Cédex 05 (France)
Demoulin V., Departement de Botanique, Université de Liège, Sart Tilman, 4000 Liège (Belgium)
Desportes Isabelle, Laboratoire d'évolution des Etres organisés, 105 Boulevard Raspail, 75006 Paris (France)
Dodge J. D., Department of Botany, Royal Holloway College, Huntersdale Callow Hill, Virginia Water, Surrey GU 25 4LN (U.K.)
Febvre Colette, Laboratoire de Protistologie marine, Station de Biologie marine, 06230 Villefranche-sur-Mer (France)
Febvre Jean, Laboratoire de Protistologie marine, Station de Biologie marine, 06230 Villefranche-sur-Mer (France)
Fernandez-Leborans G., Laboratorio de Biologia general, Facultad de Biologia, Universitad Complutense, Ciudad Universitaria, Madrid 3 (Spain)
Gadea Enrique, Departamento de Zoologia, Facultad de Biologia, Universitad de Barcelona, Gran Via 585, Barcelona 7 (Spain)
Galleron Colette, 64 rue Dutot, 75015 Paris (France)
Genermont Jean, Laboratoire de Zoologie, Université Paris Sud, 91405 Orsay (France)
Gil Rosario, C.S.I.C., Institut Immunologia y Biologia Microbiana, Velazquez 144, Madrid 6 (Spain)
Goode D., Department of Zoology, University of Maryland, College Park, MD 20742 (U.S.A.)
Gracia-Royo M. P., Departamento de Zoologia, Universitad de Barcelona, Gran via 585, Barcelona 7 (Spain)
Grain J., Groupe de Zoologie et Protistologie, Université de Clermont, Complexe Scientifique des Cézeaux, B.P. 45, 63170 Aubière (France)
Greuet C., Laboratoire de Biologie animale et Cytologie, Faculté des Sciences, 06034 Nice Cédex (France)
Grosovsky Betsey, Biology Department, Boston University, 2 Cummington Street, Boston, MA 02215 (U.S.A.)
Heath I. B., Department of Biology, York University, 4700 Keele Street, Downsview, Ontario M3J 1P3 (Canada)

Heisler J. J., Department of Zoology, University of Maryland, College Park, MD 20742 (U.S.A.)
Herzog Michel, Laboratoire Arago, 66650 Banyuls-sur-Mer (France)
Kivic P., Botany Department, University of Tennessee, Knoxville, TN 37916 (U.S.A.)
Lefort-Tran Marcelle, CNRS, Cytophysiologie de la Photosynthèse, 91190 Gif-sur-Yvette (France)
Lenk Susan E., Department of Zoology, University of Maryland, College Park, MD 20742 (U.S.A.)
Li Jing Yan, Kunming Institute of Zoology, Academia Sinica, Kunming, Yunnan Province (China)
Lipscomb D. L., The George Washington University, Department of Biological Sciences, Washington, DC 20052 (U.S.A)
Little Melvyn, German Cancer Research Center, Institute of Cell and Tumor Biology, Im Neuenheimer Feld 280, Heidelberg (F.R.G.)
Lloyd David, University College, Newport Road, Cardiff CF2 1TA (U.K.)
Ludueña R. F., Department of Biochemistry, University of Texas Health Science Center, San Antonio, TX 78284 (U.S.A.)
Lwoff André, 69 avenue de Suffren, 75007 Paris (France)
Machelon V., Laboratoire de Zoologie, Université Paris Sud, 91405 Orsay (France)
MacKerracher Lisa, York University, Department of Biology, 4700 Keele St., Downsview, Ontario M3J 1P3 (Canada)
Margulis Lynn, Biology Department, Boston University, 2 Cummington Street, Boston, MA 02215 (U.S.A.)
Melkonian Michael, Botanisches Institut, Schlossgarten 3, 4400 Münster (F.R.G.)
Metivier Christine, Laboratoire Arago, 66650 Banyuls-sur-Mer (France)
Mignot Jean-Pierre, Laboratoire de Zoologie, Complexe scientifique des Cézeaux, B.P. 45, 63170 Aubière (France)
Moestrup Ojvind, Kobenhavns Universitet, Institut für Sporeplanter, Ofarimagsgade 2 D, 1353 Kobenhavn (Denmark)
Moss Stephen, Biological Sciences, Portsmouth Polytechnic, King Henry Building, King Henry Street, Portsmouth PO1 2DY (U.K.)
Motta J., Department of Botany, University of Maryland, College Park, MD 20742 (U.S.A.)
Murrin Faye, York University, Faculty of Science, 4700 Keele Street, Downsview, Ontario M35 1P3 (Canada)
Ourisson Guy, Université Louis Pasteur, Institut de Chimie, 1 rue Blaise Pascal 67008 Strasbourg Cédex (France)
Papenheimer Alvin, Harvard University, Biological Laboratory, 16 Divinity Avenue, Cambridge, MA 02138 (U.S.A.)
Patterson D. J., Department of Zoology, University of Bristol, Woodland Road, Bristol BS8 1UG (U.K.)
Peuto Michèle, Laboratoire de Protistologie marine, Université de Nice, Parc Valrose, 06034 Nice Cédex (France)
Prevot Paul, Laboratoire Arago, 66650 Banyuls-sur-Mer (France)
Puytorac P., Biologie comparée des Protistes, Université de Clermont-Ferrand Les Cézeaux, B.P.45, 63170 Aubière (France)
Rizzo Peter, Biology Department, Texas A. & M. University, College Station, TX 77843 (U.S.A.)
Roos V. P., Department de Cytologie, Université de Zürich, Zürich (Switzerland)
Rouviere-Yaniv, Institut Pasteur, 25 rue du Dr. Roux, 75724 Paris Cédex 15 (France)
Searcy D. G., Zoology Department, University of Massachusetts, Amherst, MA 01003 (U.S.A.)
Sigee David, Department of Botany and Zoology, University of Manchester, Manchester (U.K.)
Small E., Deprtment of Zoology, University of Maryland, College Park, MD 20742 (U.S.A.)
Soyer-Gobillard Marie-Odile, Laboratoire Arago, 66650 Banyuls-sur-Mer (France)
Surek Barbara, Botanisches Institut, Schlossgarten 3, 4400 Münster (F.R.G.)
Taylor F. J. R., The University of British Columbia, 6270 University Boulevard, Vancouver, BC V6T 1W5 (Canada)
Tellez M. C., Institut d'Océanographie, Paseo Nacional, Barcelona (Spain)
Theodorides Jean, Laboratoire d'Evolution des Etres organisés, 105 Bd.Raspail, 7500 Paris (France)
Walne P. L., Department of Botany, University of Tennessee, Knoxville, TN 37916 (U.S.A.)
Westbroek Peter, Department of Biochemistry, Leiden University, Wassenaarseweg 64, 2333 AL Leiden (Holland)

FOREWORD

INTERNATIONAL SOCIETY FOR EVOLUTIONARY PROTISTOLOGY (ISEP)

5th International Meeting

Banyuls-sur-Mer, France, June 4–6, 1983

For the first time since its inception, at Boston University in June 1975[1], the Society for Evolutionary Protistology met in Europe. Under the direction of Marie-Odile Soyer-Gobillard and hosting some 70 people representing a dozen nations (Belgium, Canada, Denmark, England, France, W. Germany, The Netherlands, Poland, Scotland, Spain, Switzerland, U.S.A.) the meeting was held at Banyuls-sur-Mer in Catalunya. The 1983 ISEP met at the famed Laboratoire Arago on the Mediterranean Sea, most participants were housed in the Laboratory's newly refurbished Grand Hotel.

The previous meetings had emphasized single themes, e.g., (First) Boston, 1975 Evolution of Mitosis in Eukaryotic Microorganisms; (Second) Downsview Ontario, 1977 Criteria for Phylogeny in Protists. In spite of the fact that the third meeting, planned for Leeds, England in June of 1979, was never held some of the papers scheduled to be presented there were published in *BioSystems*, Volume 12, Numbers 1 and 2. The fourth meeting at Port Deposit, Maryland, 1981 called Conference on Cellular Evolution focused on the Evolution of Microtubules, Mitosis, Microfilaments and other Fibrillar Systems. The proceedings of this meeting were published in *BioSystems*, Volume 14, Numbers 3 and 4. This fifth meeting was planned around multiple themes: Experimental methods in studying evolution, uniformity and diversity in protistan structure, relationships between protistan phyla, relationship between nucleoid and cytoplasm in archaebacteria and nucleus and cytoplasm in eukaryotic cells, dinoflagellate chromosome organization and the origin of muticellularity. The papers from this 5th meeting are here (*Origins of Life* vol. **13**, p. 169-352 as the journal and the book) with the exception of contributions by Li-Jing Yang, D. Sigee, J. Dodge, P. Rizzo and Morris that deal with dinoflagellates. Those four promptly submitted papers appeared in *BioSystems* vol. **17**, 1984. The invited speaker at the meeting, Professor Guy Ourisson of the University of Strasbourg, introduced the protistologists to the power of organic geochemistry. He discussed studies of secondary metabolism in aiding the interpretation of phylogenies as well as the use of organic geochemical analysis in the interpretation of the fossil history of photosynthetic

1 See *BioSystems*, Volume 7, Numbers 3 and 4, Symposium on *The Evolution of Mitosis in Eukaryotic Microorganisms*.

microbes and plants. Nobel Laureate, Andre Lwoff whose book of ciliate morphogenesis[2] and techniques of ciliate cortical staining (Chatton-Lwoff technique) has provided protistological inspiration since the 1940's, was in attendance and introduced Professor Ourisson.

As emphasized by John Corliss of the University of Maryland, the protists (*sensu lato*, by which he means the *protoctists*, eukaryotic organisms exclusive of members of the Kingdoms Animalia, Plantae and Fungi) comprise a far larger and diverse group of organisms than most realize. Corliss estimates that there are more than 110,000 species of protists comprising perhaps 40 major lineages or phyla. These organisms include the 'water molds' or so-called 'motile fungi' such as *Saprolegnia* and other oomycotes that are serving as excellent material to provide the basis for understanding of mitotic movement and sexuality. This was amply demonstrated by Professor I. B. Heath and his group (F. Murrin and L. MacKerracher). A general theory of the evolution of mitotic movements was presented by U.-P. Roos from Zürich.

The polyphyly of multicellularity was demonstrated by the work of Isabelle Desportes (Paris), in work on the bizarre life cycle of the *Paramarteilia* (myxosporidians, parasite of Polychaetes) in which cells develop inside other cells of the same organism.

The use of microtubular ultrastructural patterns to assess relatedness has become apparent to everyone. Both the taxonomy of heliozoans (Colette and Jean Febvre, Villefranche-sur-Mer) and the taxonomy of ciliates (Eugene Small, College Park, Maryland) are being extensively revised. The concept of *kinetid* (*cinetid*), the unit pattern of cell cortex which is comprised of the basal structure of microtubules and microfibrils surrounding the kinetosome, is becoming crucial in the explication of the phylogeny of members of the protoctists, independent of the presence of plastids. The importance of the heterotrophic portion of the cell was elegantly pointed out by P. Kivic and P. Walne (Tennessee) in a paper that showed members of the Euglenids and Kinetoplastids (the group to which *Trypanosoma* belongs) to be far more related than euglenoids and, for example, the chlorophyte green algae.

The importance of protists in the elucidation of fundamental cell problems was demonstrated by several speakers. The presence of striated fibers involved in cell calcium regulation and movement was shown by M. Melkonian (Münster) in his work on the prasinophyte *Tetraselmis* (= *Platymonas*).

Biomineralization, for example, the intracellular production of calcium carbonate tests, is optimally studied in the haptophytes (coccolithophorids), as shown by P. Westbroek from Leiden. A fascinating hypothesis that relates light perception and directed behavior in dinoflagellates was presented by J. Dodge (Surrey) in a paper that involves a strand of microtubules (originating at the base of the longitudinal undulipodium). This strand of microtubules passes over the eyespot (in two rather different species of dinoflagellates each with different eyespot organization). Dodge

2 Problems of Morphogenesis in Ciliates: The Kinetosomes in Development, Reproduction and Evolution, 1950. New York: Wiley Publishers.

suspects that the microtubules have a direct role in the transmission of directional stimuli that bring about the phototropic response.

The uniqueness of the genetic organization of dinoflagellates was emphasized by several investigators (M. Herzog, M. O. Soyer-Gobillard, Banyuls-sur-Mer; Peter Rizzo, College Station; David Sigee, Manchester; and C. Galleron of Paris). Apparently the high quantity of hydroxymethyl uracil which replaces so much thymine in dinoflagellates appears in the DNA by means of a post-replicative mechanism. The peculiar characteristics of the dinoflagellates' nuclei strongly suggest that this group is monophyletic and has evolved independently of the other eukaryotes.

The tubulin proteins, especially beta-tubulin, comprising undulipodia are remarkably conserved in the great range of eukaryotes studied. On the other hand M. Little (Heidelberg), R. Ludueña (San Antonio) and their colleagues have shown that variations in alpha-tubulin provide fine tools for reconstructing the phylogeny of eukaryotic microbes and their relationships to animals and plants. Nonanimal alpha-tubulins, as determined by peptide digest studies, of the cytoplasm of a plant (rose) are nearly identical to the alpha-tubulins of the green algae and ciliates tested, and are very similar to cytoplasmic alpha-tubulins of the plasmodial slime mold *Physarum* and the heliozoan *Echinosphaerium*. These are in marked contrast to animal alpha-tubulins which closely resemble each other. These investigators including Andre Adoutte (Gif-sur-Yvette) hope to use tubulin sequence data and immunocytochemistry to solve the thorny problem of which protists were ancestral to animals and plants. Another approach to this classical problem came from C. Bardele (Tübingen) who showed that the details of the undulipodial necklace (membrane patterns, as revealed by freeze etching on the inside of cilia and sperm tails) show a close relatedness in all animals studied but are far more varying in protists. Perhaps by finding the protist pattern most like that of the metazoa, the extant lineage most closely related to the ancestral animals will eventually be identified.

Many ISEP members who participated in this meeting are also contributing to the *Handbook of Protoctists*. This handbook, edited by Lynn Margulis, John Corliss and David Chapman, is scheduled to be published in early 1985 by Jones and Bartlett Publishers. It will be one volume with chapters on each phylum in Kingdom Protoctista. P. Westbroek, P. Walne and P. Kivic, E. Cox, M. Melkonian, E. Small and D. Lynn, and D. Barr are some of these authors.

Due to the hard work of I. Brent Heath, the fledgling ISEP has achieved legal status as an international nonprofit scientific organization, registered in Canada. According to the by-laws a regular member of ISEP "shall be persons having an interest in the origin, evolution and phylogeny of eukaryotic organisms who have made an application to and have been accepted by the Secretary". The presidency of the Society has now passed from Christian Bardele to Professor Heath. The Secretary is Dr. Diana Lipscomb. It was decided after much discussion that the next biennial meeting will be held again in North America, at Ottawa June 10-14 1985, under the direction of Dr. Donald Barr. At that time Dr. Dennis Goode (University of Maryland), who was elected President Elect of ISEP at Banyuls, will begin his presidency.

The Banyuls meeting was beautifully organized, aid was forthcoming from several sources. Sources of money were: Centre National de la Recherche Scientifique (Secteur sciences de la vie), C.N.R.S. – PIRO (Programme interdisciplinaire pour la recherche oceanographique), Université Pierre et Marie Curie, Paris VI, Association Naturalia et Biologia (Paris). The food was remarkable and many participants greatly enjoyed their visit to Tautauvel, the small Catalunyan town. It was here, in the cave of Arago, that a fine fossil skull and other bones and teeth of *Homo erectus*, about 400,000 years old, have been found. These represent the oldest evidence for early man in Europe demonstrating the extreme desirability as living space of this magnificent, ancient part of the world.

I thank I. B. Heath, B. Dexter Dyer, Donna Mehos, and Marie-Odile Soyer-Gobillard for aid in preparing this report.

LYNN MARGULIS

THE TUBULINS OF ANIMALS, PLANTS, FUNGI AND PROTISTS IMPLICATIONS FOR METAZOAN EVOLUTION

MELVYN LITTLE

Institute of Cell and Tumor Biology, German Cancer Research Center, D-6900 Heidelberg, F.R.G.

RICHARD F. LUDUEÑA

Department of Biochemistry, University of Texas Health Science Center, San Antonio, TX 78282, U.S.A.

LOUIS C. MOREJOHN

Department of Developmental and Cell Biology, University of California, Irvine, CA 92717, U.S.A.

CLARA ASNES

Department of Biology, University of Rochester, Rochester, NY 14627, U.S.A.

and

EUGENE HOFFMAN

Department of Biology, Western Kentucky University, Bowling Green, KY 42101, U.S.A.

(Received 5 October, 1983)

Summary. α-Tubulin subunits from trout (*S. gairdneri*) sperm tails, sea urchin (*S. purpuratus*) cilia, protistan alga (*C. elongatum*) flagella and rose (Paul's Scarlet) cytoplasm have been characterized by limited proteolytic cleavage with the enzyme *Staphylococcus aureus* protease and electrophoresis of the digestion products on SDS-PAGE. The resulting patterns corresponded to either of two major types representative of animal and non-animal α-tubulins, respectively. A total of 28 α-tubulins have now been characterized by this method. They are classified in this paper according to the type of cleavage pattern generated by the enzyme *S. aureus* protease. The implications of these results for metazoan evolution are discussed.

1. Introduction

Protein sequence analysis has proved to be an excellent method for quantitating the relationships between widely different species (Dayhoff, 1978). Cytochrome c protein sequences, in particular, have been extremely useful in establishing evolutionary trees. In comparison to the abundant data from animals and bacteria, however, very few mitochondrial cytochrome c proteins have been sequenced from protists. Thus, although cytochrome c data, and to a lesser extent that of ribosomal 5s RNA and phenylalanine tRNA, indicate that animals, plants and fungi all originated at about the same time, it is not known whether these organisms arose from one group of single-celled protists or if each kingdom arose from different groups of protists. In order to investigate such problems concerning the evolution of eukaryotic organisms and their organelles it will probably be necessary to sequence other protein families which are not confined to mitochondria or chloroplasts.

Since the protein tubulin has a low mutation rate (Krauhs *et al.*, 1981; Little *et al.*,

Origins of Life **13** (1984) 169-176. 0302-1688/84/0133-0169$01.20.

1981a; Lemischka *et al.*, 1981) it is a very interesting protein from an evolutionary point of view. It is also a major component of flagella, cilia, axopodia, cytoplasm and the mitotic spindle, and is fairly easy to isolate. Moreover, although there has been a report of microtubules in prokaryotes (Margulis *et al.*, 1978), it is widely believed that the appearance of the first microtubules coincided with the evolution of eukaryotes.

In order to determine which tubulins would be interesting candidates for amino acid sequencing and to test whether peptide mapping would be a useful method for solving some evolutionary problems, we have compared the proteolytic cleavage patterns of a large number of tubulins from representative species and organelles. The enzyme *Staphylococcus aureus* protease was used under conditions of limited proteolysis (Cleveland *et al.*, 1977) to produce characteristic peptide patterns on SDS-PAGE. It was found that whereas the β-tubulin patterns were all fairly similar, the α-tubulin pattern could be used to distinguish between several families of tubulins. One major omission, however, has been a plant cytoplasmic tubulin. In this paper we have accordingly compared rose cytoplasmic tubulin to that of other tubulins. Although a comparison has already been made of rose tubulin to brain tubulin, it has not been compared to non-animal tubulins (Morejohn and Fosket, 1982). We have also enlarged our library of tubulin peptide maps by isolating and characterizing the tubulins from trout sperm, sea urchin cilia and the unicellular alga *Chloragonium elongatum*. These results and those of previous investigations have been summarized and their implications for metazoan evolution discussed.

2. Materials and Methods

Materials. Acrylamide (4 × recrystallised), N,N′-methylene-bis-acrylamide and sodium iodoacetate were from Serva (Heidelberg). *Staphylococcus aureus* V8 protease (*S. aureus* protease) was obtained from Miles Laboratories.

3. Preparations of Tubulin Samples

Tubulin from abalone (*H. rufescens*) sperm tails, sea anemone (*T. crassicornis*) sperm tails, *Tetrahymena pyriformis* cilia, rose (Paul's Scarlet, a hybrid of *Rosa multiflora* and hybrid chinas) cytoplasm and from flagella of the unicellular alga *Chlorogonium elongatum* were from previous investigations (Little *et al.*, 1982b; Morejohn and Fosket, 1982; Little *et al.*, 1983b). Sperm from the rainbow trout *Salmo gairdneri* was the kind gift of Dr. William Herschberger, the University of Washington Fisheries Research Institute, Seattle, Washington. Cilia were detached from 24 hr blastulas of the sea urchin *Strongylocentrotus purpuratus* by suspending the blastulas in 10 volumes of ice-cold hypertonic sea water (15 g of NaCl in 500 ml of filtered sea water). The cilia were purified according to Stephens (1972). Outer doublets of trout sperm tails and sea urchin cilia were isolated according to the method of Stephens (1970).

4. Isolation of α- and β-Tubulins

Tubulin was carboxymethylated according to the method of Crestfield *et al.* (1963) and the subunits were separated on 7% acrylamide slab gels using the electrophoresis system of Yang and Criddle (1970). They were located as translucent bands after incubation with 4N sodium acetate (Higgins and Dahmus, 1979). The subunits were then removed from the gel slices as previously described (Little *et al.*, 1981b) and dissolved in electrophoresis sample buffer containing 0.2% sodium dodecyl sulfate.

5. Proteolytic Cleavage Patterns

The tubulin subunits were digested with *S. aureus* protease in a 5% acrylamide stacker gel. Electrophoresis of the peptides on 15% acrylamide gels was performed according to the method of Cleveland *et al.* (1977) as previously described (Little *et al.*, 1981b). Approximately 1 μg of tubulin subunit and unless otherwise stated 0.1 μg of enzyme were applied per well. The peptides were stained with silver as described above according to the procedure of Oakley *et al.* (1980).

6. Results

The *S. aureus* proteolytic cleavage patterns of α-tubulins from trout (*S. gairdneri*) sperm tails, sea urchin (*T. crassicornis*) cilia, rose (Paul's Scarlet) cytoplasm and flagella of the

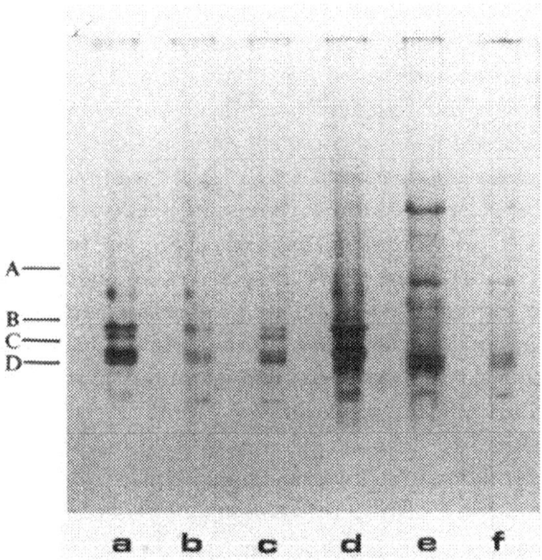

Fig. 1. *S. aureus* protease cleavage patterns of α-tubulin subunits from animal and non-animal sources. The digests were resolved by electrophoresis on 15% acrylamide gels and stained with silver. (a) Sea anemone (*T. crassicornis*) sperm tail α-tubulin; (b) sea urchin (*S. purpuratus*) ciliary α-tubulin; (c) trout (*S. gairdneri*) sperm tail α-tubulin; (d) abalone (*H. rufescens*) sperm tail α-tubulin; (e) rose (Paul's Scarlet) cytoplasmic α-tubulin; (f) *Tetrahymena pyriformis* ciliary α-tubulin.

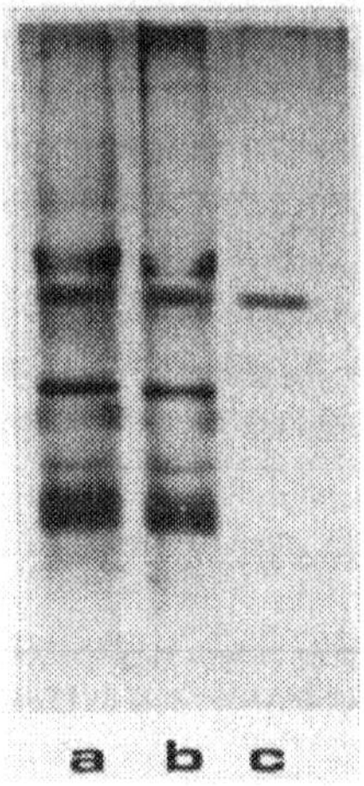

Fig. 2. *S. aureus* protease cleavage patterns of the α-tubulins from *Tetrahymena* cilia and *Chlorogonium* flagella. The digests were resolved by electrophoresis on 15% acrylamide gels and stained with silver. (a) *Tetrahymena pyriformis* ciliary α-tubulin; (b) *Chlorogonium elongatum* flagellar α-tubulin; (c) enzyme, 0.5 μg.

protistan alga *Chlorogonium elongatum* are compared to those of the well characterized α-tubulins from abalone and sea anemone sperm tails and *Tetrahymena pyriformis* cilia (Little *et al.*, 1982b) in Figure 1 and Figure 2. The patterns of the α-tubulin peptides from trout sperm tails and sea urchin cilia are very similar to that of sea anemone sperm tail α-tubulin. One small but perhaps significant difference, however, is that the sea

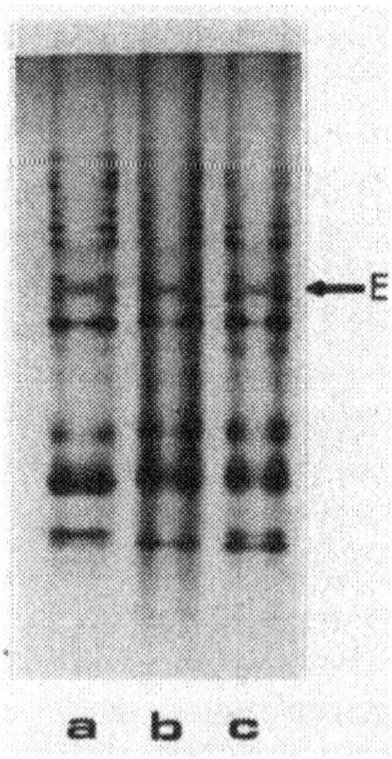

Fig. 3. *S. aureus* protease cleavage patterns of trout sperm trail and sea urchin ciliary β-tubulin subunits compared to that of sea anemone sperm tail β-tubulin. The digests were resolved by electrophoresis on 15% acrylamide gels and stained with silver (a) Trout (*S. gairdneri*) sperm tail β-tubulin; (b) sea urchin (*S. purpuratus*) ciliary β-tubulin; (c) sea anemone (*T. crassicornis*) sperm tail β-tubulin; E is enzyme.

urchin ciliary peptide in position C is present in as much quantity as the peptide in position B. This is the only case amongst the 10 animal axonemal α-tubulins so far investigated in which the peptide in position B is not present in a clearly greater quantity than the peptide in position C. The *S. aureus* protease cleavage pattern of rose cytoplasmic α-tubulin has already been shown to be different from that of brain α-tubulin (Morejohn and Fosket, 1982) and here it is clearly shown to be very similar to the α-tubulin pattern of *Tetrahymena*. Since the α-tubulin peptide pattern of *Tetrahymena* has also been shown to be almost identical to that of sperm tails from the fern *Pteridium aquilinum*, it seems that plant cytoplasmic and axonemal α-tubulins are very similar. Rose cytoplasmic β-tubulin has been shown to have an *S. aureus* protease peptide pattern like that of brain β-tubulin (Morejohn and Fosket, 1982). Indeed, the *S. aureus* protease β-tubulin peptide patterns of all species so far investigated are very similar to those of trout sperm tails, sea urchin cilia and sea anemone sperm tails shown in Figure 3, with the possible exception of axopodial β-tubulin from the heliozoan *Echinosphaerium nucleophilum* which shows some larger differences (Little *et al.*, 1983b).

7. Discussion

Twentyeight tubulins have now been characterized in this and previous investigations by one-dimensional *S. aureus* protease peptide maps (Little *et al.*, 1981b, 1982a, 1982b, 1983a, 1983b). Most of the β-tubulin peptide patterns are very similar with an even distribution of high, medium and lower molecular weight peptides on SDS-PAGE. α-Tubulin peptide patterns, however, can be divided into two main groups representative of animal and non-animal α-tubulins (Figure 4). They are summarized in Table I.

The animal axonemal α-tubulin peptide patterns closely resemble one another, although it is possible to distinguish between two sub-types designated X and Y (Little *et al.*, 1982b). In contrast, animal cytoplasmic α-tubulin peptide patterns show some

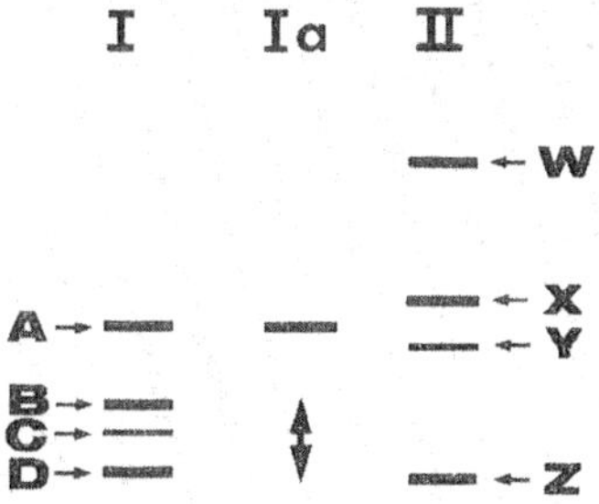

Fig. 4. Types of peptide patterns generated by the α-tubulins of animal and non-animal organisms after limited proteolytic digestion with *S. aureus* protease and electrophoresis on 15% acrylamide gels. Only the major peptides have been drawn which are consistent characteristics of the patterns. Type I: Animal axonemal α-tubulins. This type of pattern can also be resolved into one of two very similar but distinct sub-types (Little *et al.*, 1982b); Type Ia: Animal cytoplasmic α-tubulins. This type of pattern consists of a peptide in position A followed by a group of peptides in the region of the gel denoted by the double-headed arrow. Type II: Non-animal α-tubulins.

Table I

Source and type of α-tubulin

Type I	Sub-type X	Trout (*Salmo gairdneri*) sperm tail
		Sea urchin (*S. purpuratus*) cilia
		Sea urchin (*S. purpuratus*) sperm tail
		Sea urchin (*L. pictus*) sperm tail
		Clam (*M. mercenaria*) gill cilia
		Sea anemone (*T. crassicornis*) sperm tail
	Sub-type Y	Pig sperm tail
		Tunicate (*C. intestinalis*) sperm tail
		Abalone (*H. rufescens*) sperm tail
		Marine echiurid worm (*U. caupo*) sperm tail
Type Ia		Pig brain
		Bovine brain
		Bovine kidney
		[1]Pig lung
		[1]Pig blood platelets
		Dogfish brain
		Sea urchin (*S. purpuratus*) egg
		[1]Brine shrimp (*Artemia*) cytoplasm
		Squid (*L. pealei*) brain
		[2]Drosophila egg tubulin
		[3]Nematode (*A. galli*) cytoplasm
Type II		Bracken fern (*P. aquilinum*) sperm tail
		Rose (*Paul's Scarlet*) cytoplasm
		Chlamydomonas reinfardtii flagella
		Chlorogonium elongatum flagella
		Tetrahymena pyriformis cilia
		[4]Slime mold (*P. polycephalum*) cytoplasm
		Heliozoan (*E. nucleophilum*) axopodia

[1]Unpublished results.

[2]Fast migrating peptide A. See text.

[3]Dawson *et al.*, 1983.

[4]See also Clayton *et al.*, 1980.

marked differences. They are still fairly easy to recognise as animal α-tubulins, however, since they nearly all contain a peptide corresponding to axonemal peptide A in Figure 4 and the rest of the peptides have electrophoretic mobilities in the range of the axonemal peptides B, C, and D. The only exception found to this pattern of animal cytoplasmic α-tubulins is that of Drosophila egg α-tubulin (Little, 1982a), where the peptide apparently corresponding to peptide A migrates with a somewhat higher mobility.

Although all the non-animal α-tubulins so far investigated generate the type II pattern of Figure 4, some are clearly more similar than others, the corresponding peptides generated by the α-tubulins of *Tetrahymena*, *Chlamydomonas*, *Chlorogonium* and rose cytoplasm are practically identical. The *Pteridium* α-tubulin map differs only slightly from these in that the X band is a doublet instead of being a single band as in the others. Both the *Physarum* cytoplasmic and *E. nucleofilum* axopodial α-tubulin peptide patterns differ from these in that their X and Y peptides migrate further apart and their Z peptides migrate somewhat faster.

In previous papers (Little *et al*., 1982a, 1982b) we postulated that the tubulin genes of plants and protistans, including the myxomycete *P. polycephalum*, code for functionally identical tubulins. The use of the same protein for a variety of purposes imposes severe structural constraints which could account for the similar *S. aureus* protease α-tubulin cleavage patterns of these organisms. Duplications and modifications of the tubulin genes during the evolution of the metazoa may have led to the use of function-specific tubulins. The tubulin structure was then able to optimise for a diminished number of functions, thus explaining the different peptide patterns of animal and non-animal tubulins. The more conservative cleavage pattern of the animal axonemal α-tubulins compared to that of the cytoplasmic α-tubulins may reflect the more complex, rigid structure of axonemes.

Alternatively, the protistan ancestor of the metazoa may have already contained an α-tubulin generating the type I proteolytic cleavage pattern. Since cytochrome c sequence data indicate that the animals, plants and fungi diverged at about the same time (Dayhoff, 1978), an event occurring much later than the divergence of *Tetrahymena* and unicellular algae from multicellular organisms, this would mean that the metazoan protistan ancestor evolved a type I α-tubulin not long before the evolution of the metazoa. It seems more likely to us, however, that a change from a type II to a type I α-tubulin would have occurred during the period of radical genetic change accompanying metazoan evolution rather than in a protistan ancestor shortly before the advent of multicellularity.

The method of tubulin typing with *S. aureus* protease is a relatively simple method of finding out just when the change occurred from the type of tubulin found in plants, protists and fungi to that found in the metazoa. If for example, the change occurred in a protist group, descendants may be discovered which contain the type I tubulin. These would be the 'missing links' in the evolution of protists to metazoa! It would also be very interesting to look at the tubulin from a group of metazoa thought to have diverged early from the rest of the metazoa, the sponges. If they contain the type I α-tubulin, then this type of tubulin would have been present in metazoan organisms before the evolution of specialized nerve and muscle cells.

Acknowledgements

We are very grateful to Ellen Heusterberg for her skillful technical assistance. This work was supported by a Robert A. Welch foundation grant AQ-726 to R.F.L., a Western Kentucky University Faculty grant to E. H. and a NATO research grant award Nr. 1677 to M.L.

References

Clayton, L., Quinlan, R. A., Roobol, A., Pogson, C. I., and Gull, K.: 1980, A Comparison of Tubulins from Mammalian Brain and Physarum Polycephalum Using SDS-poly-acrylamide Gel Electrophoresis and Peptide Mapping, *FEBS Lett.* **115**, 301.

Cleveland, D. W., Fischer, S. G., Kirschner, M. W., and Laemmli, U. K.: 1977, Peptide Mapping by Limited Proteolysis in Sodium Dodecyl Sulfate and Analysis by Gel Electrophoresis, *J. Biol. Chem.* **252**, 1102.

Crestfield, A. M., Moore, S., and Stein, W.H.: 1963, The Preparation and Enzymatic Hydrolysis of Reduced and S-carboxymethylated Proteins, *J. Biol. Chem.* **238**, 622.

Dawson, P. J., Gutteridge, W. E., and Gull, K.: 1983, Purification and Characterization of Tubulin from the Parasitic Nematode, *Ascaridia galli, Mol. Biochem. Parasitol.* **7**, 267.

Dayhoff, M. O.: 1978, Atlas of Protein Sequences and Structure (National Biomedical Research Foundation, Washington D.C.) Vol. 5, Suppl. 3.

Higgins, R. C. and Dahmus, M. E.: 1979, Rapid Visualization of Protein Bands in Preparative SDS-polyacrylamide Gels, *Anal. Biochem.* **93**, 257.

Krauhs, E., Little, M., Kempf, T., Hofer-Warbinek, R., Ade, W., and Ponstingl, H.: 1981, Complete Amino Acid Sequence of β-tubulin from Porcine Brain., *Proc. Natl. Acad. Sci. U.S.A.* **78**, 4156.

Lemischka, I. R., Farmer, S., Racaniello, V. R., and Sharp, P. A.: 1981, Nucleotide Sequence and Evolution of a Mammalian α-tubulin Messenger RNA, *J. Mol. Biol.* **151**, 101.

Little, M., Krauhs, E., and Ponsting, H.: 1981a, Tubulin Sequence Conservation, *BioSystems* **14**, 239.

Little, M., Röhricht, C, and Schroeter, D.: 1983a, Pig Sperm Tubulin: Its Extraction and Characterization, *Exp. Cell Res.* **147**, 15.

Little, M., Krauhs, E., Ponstingl, H., Ludueña, R. F., and Rice, N.: 1982a, Tubulin Subunits from Higher and Lower Eukaryotes – Structural Comparison, in P. Cappuccinelli, and N.R. Morris (eds.), *Microtubules and Microorganisms*, pp. 1–14, Publ. Marcel Dekker, New York and Basel.

Little, M., Ludueña, Keenan, R., and Asnes, C. F.: 1982b, Tubulin Evolution: Two Major Types of α-tubulin, *J. Mol. Evol.* **19**, 80.

Little, M., Ludueña, R. F., Langford, G. M., Asnes, C. F., and Farrell, K.: 1981b, Comparison of Proteolytic Cleavage Patterns of α- and β-tubulins from Taxonomically Distant Species, *J. Mol. Biol.* **149**, 95.

Little, M., Quinlan, R. A., Ludueña, R. F., and Hoffman, E. J.: 1983b, Identification and Characterization of Axopodial Tubulins from *Echinosphaerium nucleophilum*, *Eur. J. Cell Biol.* **31**, 55.

Margulis, L., To, L., and Chase, D.: 1978, Microtubules in Prokaryotes, *Science* **200**, 1118.

Morejohn, L. C. and Fosket, D. E.: 1982, Higher Plant Tubulin Identified by Self-Assembly into Microtubules in vitro, *Nature* **297**, 426.

Oakley, B. R., Kirsch, D. R., and Morris, N. R.: 1980, A Simplified Ultra-Sensitive Silver Stain for Detecting Proteins in Polyacrylamide Gels, *Anal. Bioch.* **105**, 361.

Stephens, R. E.: 1970, Thermal Fractionation of Outer Fiber Doublet Micro-Tubules into A- and B-subfiber Components: A- and B-tubulin, *J. Mol. Biol.* **47**, 353.

Stephens, R. E.: 1972, Studies on the Development of the Sea Urchin *Strongylocentrotus droebachiensis*. III. Embryonic Synthesis of Ciliary Proteins, *Biol. Bull.* **142**, 489.

Yang, S. and Criddle, R. S.: 1970, In vitro Biosynthesis of Membrane Proteins in Isolated Mitochondria from Saccharomyces Carlsbergensis, *Biochemistry* **9**, 3063.

TUBULIN EVOLUTION: AN ELECTROPHORETIC AND IMMUNOLOGICAL ANALYSIS

A. ADOUTTE, M. CLAISSE, and J. CANCE
Centre de Génétique Moléculaire du CNRS, 91190 Gif-sur-Yvette, France

(Received 12 December, 1983)

Abstract. This paper summarizes a survey of the electrophoretic behavior of the tubulins of 23 species (mostly protists) as well as their reactivity towards 4 anti-tubulin antibodies (raised against two ciliate tubulins and two vertebrate ones). Some generalizations concerning the relative migration rates of α VS β tubulin could be made, in particular the α/β inversion, first described in *Physarum* was extended to several ciliates. Anti-vertebrate tubulin antibodies displayed a very broad spectrum of reactions, reacting with virtually all the species tested. They appear to correspond to auto-antibodies no exclusively directed against species specific determinants. In contrast, the two anti-ciliate tubulin antibodies displayed a narrow species specificity reacting only with a limited subset of protists. They were shown to be specific for a small number of immunological determinants present on ciliate tubulins. This allowed a rough evaluation of evolutionary relatedness between the various groups of protists analyzed. The results are discussed within the framework of a number of published phyllogenies and shown to be in striking agreement with some of the schemes.

0. Introduction

The α- and β-tubulins are among the most highly conserved proteins during evolution in terms of amino acid sequence and structure. This is specially true among vertebrates (see e.g. Wilde *et al.*, 1982). However, work from several laboratories has disclosed substantial differences between the tubulins of lower and higher eukaryotes in terms of electrophoretic mobilities and peptide maps (Little *et al.*, 1981; Little *et al.*, 1982; Clayton *et al.*, 1980; Adoutte *et al.*, 1982). As previously established for *Physarum*, the tubulins of *Paramecium* migrate atypically both in 1 and 2D gels with respect to those of vertebrates: the β-tubulins of *Paramecium* and *Physarum* approximately co-migrate with that of vertebrates while the α-tubulin migrates *ahead* of β-tubulin. In addition, the peptide map of the α-subunit obtained according to the method of Cleveland *et al.* (1977) is quite different from that of vertebrates. In parallel with these electrophoretic differences we had observed that an antibody raised against *Paramecium* axonemal tubulin (Cohen *et al.*, 1982) displayed a narrow species specificity, i.e. it reacted strongly with *Paramecium* tubulins both in immunoblotting and in immunocytology experiments but did not react with vertebrate tubulins (Adoutte *et al.*, 1982).

In this work we briefly report that we have aimed at understanding the basis of this narrow immunological specificity, and at extending the results to a large number of species. We have found that this approach can lead to a number of evolutionarily significant observations.

Origins of Life **13** (1984) 177-182. 0302-1688/84/0133-0177$00.90.

			Anti-Paramecium (rabbit)		Anti-Tetrahymena (rabbit)	Anti-Turkey (rabbit)		Anti-pig (sheep)		Data provided by
Ciliates	Holotrichs	Paramecium tetraurelia	+ + +	***	+ + +	+ +/+ + +	**⁰	+(α)	**⁰	Adoutte *et al*. and Cohen
		Tetrahymena pyriformis	+ + +		+ + +					Adoutte *et al*.
	Hypotrichs	Euplotes eurystomus	+ + +	***		+ + +	**⁰			Fleury
		Stylonichia mytilus	+ + +	***						Helftenbein & Fleury
Sporozoa	Coccidia	Sarcocystis tenella	+ +		+ +	+ + +				Adoutte *et al*; and
		Toxoplasma gondii	+ +		+ +	+ + +				Porchet and Delacourte
'Flagellates'	Dinoflagellates	Amphidinium carteri	+ + +			+ +				Adoutte and Matthys
		Noctiluca miliaris		***						Metivier and Soyer
		Prorocentrum micans		***						Metivier and Soyer
	Chrysomonads	Ochromonas danica		***						Adoutte and Brugerolle
	Cryptomonads	Chilomonas paramecium		***						Adoutte and Brugerolle
	Chlorophytes (Volvocales)	Chlamydomonas reinhardi	+/+ +		+ +	+/+ +				Adoutte *et al*.
	Euglenomonads	Menoidium bibacillatum			*/**					Adoutte and Brugerolle
		Distigma proteus		*/**						Adoutte and Brugerolle
		Entosyphon sulcatum		*/**						Adoutte and Brugerolle
	Kinetoplastids	Trypanosoma brucei	–		–	+ +/+ + +				Adoutte and Gallo
	Trichomonads	Trichomonas vaginalis		0						Adoutte and Brugerolle
Rhizopodes	Radiolarians	Collozoum (and others)		0						Goode
Myxomycetes		Physarum polycephalum	–/ε(β)	0	–	+(β)		+/+ +		Adoutte *et al*.
Vertebrates	Birds	Chicken	–		–	+ + +		+ + +		Adoutte *et al*.
	Mammals	Mouse	–	0	–	+ + +		+ + +	***	Adoutte *et al*.; and Hill
		Rat	–		–	+ + +		+ + +		Adoutte *et al*.
		Pig	–		–	+ + +		+ + +		Adoutte *et al*.

Summary of the immunological reactivity tests of 23 species towards anti-tubulin antibodies. The tests were carried out either by blotting (+ and – signs) or by immunocytochemistry (* and 0 signs) using the methods described in Cohen *et al*. (1982). + + + (or ***) refer to reactions of maximal intensity, + +, + and to reactions of average, weak or very weak intensity and – (or 0) to fully negative reactions. When α and β tubulins responded differently, the positive band is indicated in brackets. The ⁰ sign following some * signs refers to the fact that in those tests, microtubules within the cell body reacted much more strongly than those of cilia.

1. Electrophoretic and Immunological Properties of Protozoan Tubulins

The species that have been analyzed are listed in the table. The electrophoretic behavior of their tubulins on 1-dimensional SDS gels can be summarized as follows: all the protozoa (except *Chlamydomonas* which is known from the work of other groups not to display the α/β inversion) display a faster migration of their tubulins than that of vertebrates; within this general rule there are, however, differences from one species to another. *Physarum*, for instance, displays the most striking difference with respect to vertebrates, its α-tubulin migrating much faster than the fastest of the two vertebrate bands, yielding an apparent molecular weight difference of 3 to 4000 daltons. We have confirmed the α/β inversion in this species (first reported by Clayton *et al.*, 1980) and have extended this observation to *Paramecium* and *Tetrahymena*. Gallo (personal communication) has made the same observation in *Trypanosoma brucei*. Since these organisms (a myxomycete, two ciliates and a parasitic flagellate) belong to quite distinct groups it is therefore possible that the α/β inversion of tubulins under specific electrophoretic conditions might be a common property of a large spectrum of lower eucaryotes. Lack of totally or partially purified tubulins from the other species listed has prevented us from extending this conclusion.

The immunological results have been summarized in the table. The first important point emerging from the data is that on one hand anti-*Paramecium* and anti-*Tetrahymena* tubulin antibodies behave identically and react only with a subset of the listed species (a subgroup of the protists) while anti-turkey and anti-pig tubulin antibodies on the other hand, behave similarly and react with the tubulins of *all* the species tested (from vertebrates to protists). These results can be explained as follows. Because of the very high degree of conservation of the amino acid sequence of tubulins among vertebrates, the tubulins of a given vertebrate species are not 'recognized' as foreign proteins by rabbits (or sheep). It is only through hyper-immunisation that immune sera can be obtained (with difficulty and usually with a low titer; see e.g. Osborn and Weber (1982) and Karsenti *et al.* (1977)). The antibodies that are obtained under these conditions can be assimilated to auto-antibodies not directed towards 'foreign', species-specific determinants but rather against 'unselected' accessible determinants. Because these determinants are not species specific they may be shared by a very large number of species, hence the broad reaction spectrum of the antibodies.

In contrast, the tubulins of protozoa would display sufficient differences from those of vertebrates (as suggested by their electrophoretic and peptidic differences) to be recognized as foreign molecules and elicit the production of specific antibodies. Such antibodies will therefore only react with tubulins from species sharing at least some of the same determinants, hence the narrowness of the reaction spectrum.

In the framework of this hypothesis it could be anticipated that the anti-vertebrate tubulin antibodies would react with a larger number of peptides of tubulins than the anti-protozoa tubulin antibodies. To test this prediction we have reacted the various antibodies over blots of *peptide maps* of purified α- and β-tubulins from *Paramecium* and pig. The maps were generated in SDS polyacrylamide gels according to the method of Cleveland *et al.* (1977). The results were quite clear in that the two anti-vertebrate antibodies reacted with several peptides (both of *Paramecium* and pig) while the two anti-protozoa antibodies reacted only with a single band of α- as well β-tubulin from *Paramecium*. Furthermore the reaction of the anti-protozoan antibodies occurred only on a partially digested peptide obtained by a very mild protease digestion: when digestion was carried out to yield the usual peptide pattern, no reaction occurred. This further suggests that the antibodies react with a single antigenic determinant in α- as well as β-tubulin which is cleaved during digestion. All together these results are in good agreement with the hypothesis and provide an explanation for the seemingly contradictory results obtained with the two types of anti-tubulin antibodies.

2. Immunological Cross Reaction and Phylogeny

The fact that the anti-*Paramecium* tubulin antibody reacted with a very limited number of antigenic determinants now provided a tool for ascertaining the presence of these determinants in a variety of species. The results of the current survey are summarized in the table. The good internal consistency of the data, i.e. the fact that the different species or suborders of a given class of protists, respond identically illustrates the validity of the approach and allow us to draw some evolutionary conclusions.

1. The antigenic determinants recognized by the anti-*Paramecium* and the anti-*Tetrahymena* (two ciliated protozoa) tubulin antibodies are probably very 'ancient' in evolutionary terms since they are also found (with a high intensity reaction) in two suborders of dinoflagellates. These organisms are often considered as *primitive* eukaryotes because of the very peculiar organization of their genetic material (reviewed by Herzog *et al.*, this issue). If a major branching occurred early in the history of eucaryotes, leading to the typical eucaryotes, on one hand, and to the dinoflagellates on the other, as suggested by Herzog *et al.* (this issue), it can be assumed that molecular features that are shared by the two lines were already present in the ancestral organisms that gave rise to the branching. Alternatively, as pointed out by Dr Cavalier-Smith, while it can be agreed upon that the immunological determinants under analysis were present in the common ancestor to dinoflagellates and ciliates, this common ancestor may not be located at the root of the entire eucaryote tree but rather at the base of those eucaryotes with mitochondria (see Cavalier-Smith, 1983). It is interesting in this respect that *Trichomonas* was negative in our test.

2. If the former idea is accepted, weak or negative responses to the antibodies by eukaryotic tubulins may indicate *partial* or *total loss* of the reactive determinants or a variety of secondary modifications preventing immunological recognition. These responses can then be taken as a very rough indication of the degree of phylogenetic divergence. It is interesting in this respect that while a Chrysomonad and a Cryptomonad yield a strong immunocytochemical reaction the three Euglenomonads as well as the Chlorophyte are only weakly positive and a *Trypanosoma* is fully negative. These results are in agreement with the scheme of phylogenetic relationships among flagellates proposed by Dodge in 1979 on the basis of a number of ultrastructural and

biochemical characters. For instance, starting from an ancestral flagellate, Dodge postulates the early divergence of a line leading to Euglenomonads and Chlorophytes. Within this line he suggests very early divergence of a line leading to Trypanosomes. Thus the whole group of organisms yielding weak or negative reactions are grouped on the phylogenetic scheme. Similarly, those yielding positive reactions belong to two other lines, one diverging very early towards the Dinoflagellates, the other comprising the Cryptomonads and the Chrysomonads (among others). Similar groupings, consistent with the data presented in this paper, can be found in Taylor's (1978) as well as in Cavalier-Smith's (1978, 1981) schemes of protist phylogeny such as on the one hand the proximity of Ciliates, Dinoflagellates and Cryptomonads (all positive in our test) and the proximity of Euglenoids and Trypanosomes on the other hand (weakly positive or negative in our test). Trichomonads (negative in our test) are distantly related to all other protists in several schemes. The fact that Trypanosomes did not react with the anti-protozoa tubulin antibodies is also in line with the recently established genomic organization of the α- and β-tubulin genes in these organisms which is most unusual, even among protists (Thomashow *et al.*, 1983).

These examples illustrate the potential usefulness of such species specific antibodies in ascertaining evolutionary relationships. In particular some specific predictions of the reactivity of a number of groups can now be made and confronted with the available phylogenetic schemes (for example on Cavalier-Smith's explicit scheme of 1981). The immunocytological and the immunochemical approach being fairly easy to carry out experimentally it can be extended to a large number of species without elaborate biochemical purifications. In this respect, it will be particularly interesting to check whether any lower metazoan reacts with these antibodies. While the data obtained are not at the level of primary genetic information (i.e. nucleic acid or protein sequences) they do reflect this level since they rest on divergence in amino acid sequences of two highly conserved proteins to yield 'immunological determinants'. In conclusion, these data and those of Little *et al.*, point out to the usefulness of tubulins as phylogenetic markers.

Acknowledgments

We are grateful to the following colleagues for having generously provided sera, biological samples or unpublished results obtained with the anti-paramecium tubulin serum: I. Barahona, J. Beisson, G. Brugerolle, J. Cohen, A. Delacourte, G. Filiatreau, A. Fleury, J. M. Gallo, D. Goode, E. Helftenbein, A. M. Hill, E. Matthys, C. Métivier, D. Pantaloni, E. Porchet, M-O. Soyer, and M. Wright. This work was supported by a grant from the Ministère de l'Industrie et de la Recherche and the Ligue Nationale Française contre le Cancer.

References

Adoutte, A., Cohen, J., Hill, A. M., Pantaloni, D., and Beisson, J.: 1982, 'Biochemical and Immunological Comparison of *Paramecium* and Vertebrate Tubulins, *Biol. Cell* **45**, 265a.

Cavalier-Smith, T.: 1978, 'The Evolutionary Origin and Phylogeny of Microtubules, Mitotic Spindles and Eucaryote Flagella', *BioSystems* **10**, 93–114.

Cavalier-Smith, T.: 1981, 'The Origin and Early Evolution of the Eucaryotic Cell', in Carlile, Collins and Moseley (eds), *Molecular and Cellular Aspects of Microbial evolution*, Society for General Microbiology Ltd. Symposium 32, Cambridge University Press.

Cavalier-Smith, T.: 1983, 'Symbiosis and Eucaryote Phylogeny: The Simultaneous Diversification of Cilia, Mitochondria And Chloroplasts', in W. Schwemmler and M. Schenk. (eds), *Endocytobiology II*, de Gruyter, Berlin, in press.

Clayton, L., Quinlan, R. A., Roobol, A., Pogson, C. I., and Gull, K.: 1980, 'A Comparison of Tubulins from Mammalian Brain and *Physarum polycephalum* Using SDS-polyacrylamide Gel Electrophoresis and Peptide Mapping, *FEBS Lett.* **115**, 301-305.

Cleveland, D. W., Fischer, S. G., Kirschner, M. W., and Laemmli, U. K.: 1977, 'Peptide Mapping by Limited Proteolysis in Sodium Dodecyl Sulfate and Analysis by Gel Electrophoresis', *J. Biol. Chem.* **252**, 1102–1106.

Cohen, J., Adoutte, A., Grandchamp, S., Houdebine, L. M., and Beisson J.: 1982, 'Immunocytochemical Study of Microtubular Structures Throughout the Cell Cycle of *Paramecium*, *Biol. Cell* **44**, 35–44.

Dodge, J. D.: 1979, 'The Phytoflagellates: Fine Structure and Phylogeny', in M. Levandowsky and S. H. Hutner (eds), *Biochemistry and Physiology of Protozoa*, Acad. Press, New-York, pp. 7–57.

Herzog, M., von Boletzky, S., and Soyer, M. O.: 1984 'Ultrastructural and Biochemical Nuclear Aspects of Eukaryote Classification: Independant Evolution of the Dinoflagellates as a Sister Group of the Actual Eukaryotes?', *Origins of Life* **13** (3) (this issue).

Karsenti, E., Guilbert, B., Bornens, M., and Avrameas, S.: 1977, *Proc. Natl. Acad. Sci. U.S.A.* **74**, 3997–4001.

Little, M., Luduena, R. F., Keenan, R., and Asnes, C. F.: 1982, 'Tubulin Evolution: Two Major Types of α-tubulin', *J. Mol. Evol.* **19**, 80–86.

Little, M., Luduena, R. F., Langford, G. M., Asnes, C. F., and Farell, K.: 1981, 'Comparison of Proteolytic Cleavage Patterns of α-tubulins and β-tubulins from Taxomomically Distant Species', *J. Mol. Biol.* **149**, 95–107.

Osborn, M. and Weber, K.: 1982, 'Immunofluorescence and Immunocytochemical Procedures with Affinity Purified Antibodies: Tubulin-Containing Structures', in L. Wilson (ed.), *Methods in Cell Biology*, Acad. Press, New York, **24**, 97–132.

Taylor, F. J. R.: 1978, 'Problems in the Development of an Explicit Hypothetical Phylogeny of the Lower Eucaryotes', *BioSystems* **10**, 67–89.

Thomashow, L. S., Milhausen, M., Rutter, W. J., and Agabian, N.: 1983, 'Tubulin Genes Are Tandemly Linked and Clustered in the Genome of *Trypanosoma brucei*', *Cell* **32**, 35–43.

Wilde, C. D., Crowther, C. E., Cripe, T. P., Gwo-Shu, M., and Cowan, N. J.: 1982, 'Evidence that a Human β-tubulin Pseudogene is Derived from Its Corresponding mRNA, *Nature* **297**, 83–84.

FROM PROTO-MITOSIS TO MITOSIS – AN ALTERNATIVE HYPOTHESIS ON THE ORIGIN AND EVOLUTION OF THE MITOTIC SPINDLE

U.-P. ROOS

Institute of Plant Biology, University of Zurich, Zollikerstrasse 107, CH-8008 Zurich, Switzerland

(Received 5 October, 1983)

Abstract. Based on the assumption that the ancestral proto-eukaryote evolved from an ameboid prokaryote I propose the hypothesis that nuclear division of the proto-eukaryote was effected by the same system of contractile filaments it used for ameboid movement and cytosis. When the nuclear membranes evolved from the cell membrane, contractile filaments remained associated with them. The attachment site of the genome in the nuclear envelope was linked to the cell membrane by specialized contractile filaments. During proto-mitosis, i.e., nuclear and cell division of the proto-eukaryote, these filaments performed segregation of the chromosomes, whereas others constricted and cleaved the nucleus and the mother cell. When microtubules (MTs) had evolved in the cytoplasm, they also became engaged in nuclear division. Initially, an extranuclear bundle of MTs assisted chromosome segregation by establishing a defined axis. The evolutionary tendency then was towards an increasingly important role for MTs. Spindle pole bodies (SPBs) developed from the chromosomal attachment sites in the nuclear envelope and organized an extranuclear central spindle. The chromosomes remained attached to the SPBs during nuclear division. In a subsequent step the spindle became permanently lodged inside the nucleus. Chromosomes detached from the SPBs and acquired kinetochores and kinetochore-MTs. At first, this spindle segregated chromosomes by elongation, the kinetochore-MTs playing the role of static anchors. Later, spindle elongation was supplemented by poleward movement of the chromosomes. When dissolution of the nuclear envelope at the beginning of mitosis became a permanent feature, the open spindle of higher eukaryotes was born.

1. Introduction

The origin and evolution of the eukaryotic nucleus and of the mitotic spindle are intriguing subjects that have inspired many authors to publish imaginative hypotheses or models. Although differing in detail, models for the origin of the nucleus have in common the assumption that a prokaryotic ancestor provided the nucleocytoplasm and that the nucleus was formed as a consequence of internalization of the genome and subsequent compartimentation (Allsop, 1969; Cavalier-Smith, 1975, 1978, 1981; Heath, 1980; Margulis, 1970, 1981; Pickett-Heaps, 1974, 1975; Schwemmler, 1979; Taylor, 1976; Uzzell and Spolsky, 1974).

More diverse are the scenarios for the origin and evolution of mitosis. The "Serial Endosymbiont Theory" (SET: Taylor, 1974) predicts that microtubules (MTs), and possibly MT complexes homologous to the axonemes of extant flagellated eukaryotes, originated in spirochete-like prokaryotes and that, following endosymbiosis, the mitotic apparatus evolved through their progressive differentiation, either before or after the nucleus was completely formed (Margulis, 1970, 1981; Schwemmler, 1979). In contrast, non-symbiont models are based on the assumption that MTs and the mitotic spindle evolved endogenously (Cavalier-Smith, 1975, 1981; Heath, 1980; Taylor,

Origins of Life **13** (1984) 183-193. 0302-1688/84/0133-0183$01.65.

1976). Other authors have left open the question about the symbiotic or non-symbiotic origin of MTs (Kubai, 1975; Pickett-Heaps, 1974, 1975).

Cavalier-Smith (1978) contested the homology between the axial filaments of spirochetes and eukaryotic MTs and discussed arguments that speak against the evolution of flagella with a 9 + 2 axoneme in prokaryotes as proposed by the SET (Margulis, 1981; Schwemmler, 1979). Reports that prokaryotic tubules are biochemically related to eukaryotic MTs (Fracek, 1981; Margulis *et al.*, 1978, 1981) have not yet been supported by convincing evidence. It is furthermore unlikely that a prokaryotic genome, extinct or extant, could code for all the polypeptides that make up the complex eukaryotic flagella (Cavalier-Smith, 1978).

Authors who have written essays on cell evolution agree that ameboid movement is more ancient than flagellar movement, which implies that a system of cell motility perhaps similar to the eukaryotic actin-myosin system often cited in this context existed before the evolution of motility mediated by MTs (Cavalier-Smith, 1975; Heath, 1980; Margulis, 1981; Schwemmler, 1979; but see Cavalier-Smith, 1978, for a different point of view). Most previously published hypotheses on the origin of mitosis considered the evolution of the second motility system, viz. the one based on MTs, a prerequisite for the division of the primitive eukaryotic nucleus. An exception is the intriguing intranuclear 'actin spindle' that Heath (1980) recently postulated. The mechanism of nuclear division I propose in the following paragraphs for the ancestral eukaryote operated without MTs as well; the system of contractile filaments that performed all the other motile functions also segregated the genomes and divided the nucleus and the cell.

2. Fundamental Considerations

"The most basic function of an apparatus for cell division is to ensure that the replicated genetic material is equally and exactly divided into two genetically identical entities and that these are distributed in such a way that each daughter cell receives one complement of this genetic material" (condensed from Pickett-Heaps, 1975). In extant eubacteria the functions of genome 'division' and genome distribution (segregation) are apparently associated (Jacob *et al.*, 1963; Ryter, 1968), but in eukaryotes they are dissociated, for the division apparatus is involved in the *segregation* of the daughter genomes, but not in the division of the genetic material into two genetically identical entities.

In its most simple form any mechanism that separates daughter genomes by a minimal distance and ensures that the plane of division runs between them (Figure 1a) would fulfill the basic function defined by Pickett-Heaps (1975), but its reliability would probably not be sufficient to exclude occasional misdivisions. Reliability can be improved in two ways: (i) by increasing the distance between the genomes, and (ii) by moving them along a defined axis. A combination of these two factors would obviously decrease the rate of failure even further (Figure 1b). Concrete examples of such improvements are: anchoring the genomes statically in pole structures that an axial division apparatus moves apart (Figure 1c), or supplementing such a mechanism with dynamic anchorings that move the genomes poleward (Figure 1d).

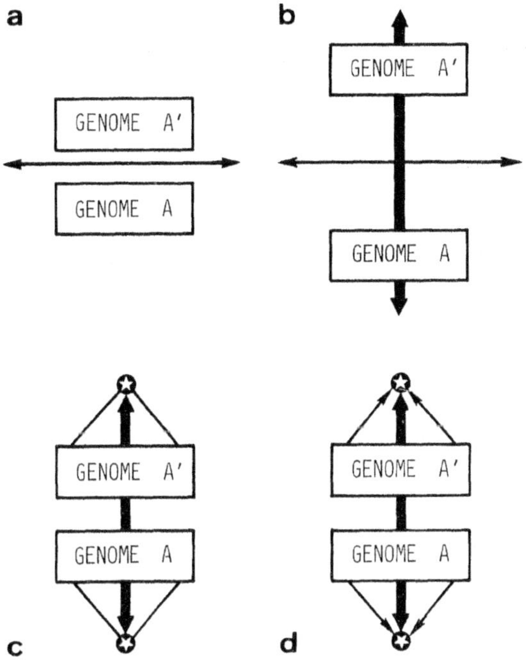

Fig. 1. (a) The basic requirements for genome segregation are a minimal distance between daughter genomes and a plane of division that separates them. (b) The rate of faultless segregation is greater if the daughter genomes separate along a defined axis and if the distance between them increases. This can be accomplished by anchoring the daughter genomes statically to pole structures that the division mechanism drives apart (c). An even more efficient and reliable mechanism involves, in addition to the elongation mechanism, the active movement of the genomes towards the poles (d).

Elements of these fundamental considerations bear on the hypothesis I present in the following paragraphs. It is based on the idea that proto-mitosis, a type of nuclear division without MTs, was a possible forerunner of mitosis and that the tendency of spindle evolution was towards greater reliability of the segregation mechanism.

3. Origin of the Nucleus

The starting point of the model (Figure 2a) is a naked prokaryotic cell (cf. Margulis, 1981) that possessed a motility system based on the interaction between filaments. This contractile system allowed the cell to move analogously to modern amebae (Allen and Taylor, 1975) and to phagocytize its prey. It could have been similar to the actin-myosin system of extant eukaryotes, as Cavalier-Smith (1975) suggested, or a forerunner of it, but it could also have been composed quite differently, yet fulfilling a similar function. Whether actin-like proteins occur in extant prokaryotes is still controversial (Gilbert and Fulton, 1982; Maniloff, 1981). The archaebacterium *Thermoplasma acidophilum*, which many authors consider a possible relative of the ancestor of the eukaryotic nucleo-cytoplasm (review: Gray and Doolittle, 1982), contains 6nm-filaments (Searcy *et al.*, 1981). Cytoplasmic extracts from this prokaryote are contractile and also contain microfilaments, which, however, are apparently unlike eukaryotic actin, for they do not bind heavy meromyosin (Searcy *et al.*, 1981). Thus, it is at least possible that there are contractile systems in prokaryotic cells that are different from the actin-myosin system and may well be evolutionary relics from the time before the appearance of the first eukaryotic cells. Among modern eukaryotes there are also organisms that have motility systems not based on an interaction between actin and myosin (cf. Cachon and Cachon, 1981). Examples are the spasmoneme of stalked ciliates (Amos, 1975; Routledge, 1978), the axopodial motility system of *Stycholonche zanclea* (Cachon *et al.*, 1977), and the ameboid movement of nematode sperm (Nelson *et al.*, 1982). Even more relevant in this context are the recent findings of Hobbs *et al.* (1983) that mitosis and contractility in the ciliate *Spirostomum teres* apparently do not involve actin. As one reviewer commented, actin may not be as ancient a protein as was generally assumed to date. In my opinion, motility based on actin and myosin has been overemphasized and it may be more profitable to examine prokaryotes (and eukaryotes, notably the lower ones) for motile and contractile systems based on other filament-forming proteins.

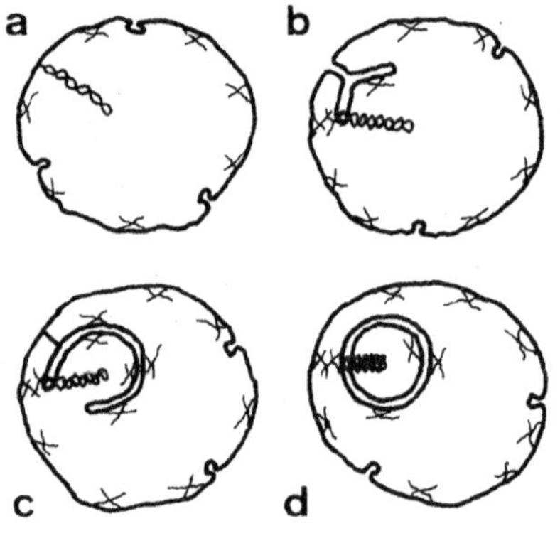

Fig. 2. Evolution of a proto-eukaryotic cell (Fig. 2d) from an ameboid prokaryotic ancestor (Fig. 2a). A result of the formation of the nucleus (Fig. 2b, c) is that filaments (shown by x's) of the primitive contractile system remain also associated with the membranes of the nuclear envelope. Specialized cytoplasmic filaments connect the attachment point of the genome to the cell membrane (Fig. 2d). See text for a detailed explanation.

Cavalier-Smith (1975, 1978) and Taylor (1976) assumed that eukaryotes evolved from a walled prokaryotic ancestor and that the loss of the cell wall was a secondary event. I consider a naked prokaryote ancestor to have been more primitive than walled prokaryotes because the elaboration of a cell wall requires additional genetic information which the naked cell had not yet acquired (extant eubacterial mycoplasmas are probably not primitive, but highly evolved prokaryotes derived from more complex eubacteria; Maniloff, 1981). Also, a naked, ameboid prokaryote presumably had a selective advantage over walled prokaryotes, because it could both move actively and phagocytize its prey.

The genome (DNA) of this naked prokaryote was bound at one point to the cell membrane (Figure 2a) in a manner analogous to extant prokaryotes (Jacob *et al.*, 1963). The genome and its attachment point subsequently became internalized by membrane invagination (Uzzell and Spolsky, 1974) or endocytosis (Cavalier-Smith, 1975), but a persistent connection between the attachment point and the cell membrane was maintained by the system of contractile filaments (Figure 2b). The infolded double membrane eventually closed around the genome and evolved to the nuclear envelope of the proto-eukaryote (Figure 2c, d). Obviously, filaments remained associated with the inner and outer membranes of the nuclear envelope, a condition with further consequences for the evolution of nuclear division (see below). As the genome became more complex it evolved into one or a few chromosomes (Allsop, 1969; Figure 2c, d) that were anchored collectively in the nuclear envelope by an organelle that had concomitantly differentiated from the simple attachment point of the original genome. This nucleus-associated organelle (NAO; Girbardt and Hädrich, 1975) within the nuclear envelope also ensured a connection with the cell membrane via the filament system (Figure 2d).

4. Proto-Mitosis

Proto-mitosis is by definition a form of nuclear division that preceded the appearance of MTs and the evolution of the mitotic spindle (Figure 3). Proto-mitosis began with the duplication of the chromosomes and of the NAO, the descendants of which were connected separately to the cell membrane via specialized parts of the contractile system (Figure 3a). Controlled directional action of the motility system then moved the NAOs to opposite points of the nucleus, as the latter and the cell began to elongate (Figure 3b). This concept, based on the *contractility of extranuclear elements*, is different from that of Heath (1980) who postulated that the *intranuclear* NAOs were moved apart by a *pushing action* of the '*actin spindle*'. Such a mechanism seems rather unusual, though not impossible, as exemplified by the actin bundle providing the pushing force for the extension of the acrosomal process during fertilization in some echinoderms (Tilney, 1978). Nevertheless, all other motility processes involving actin depend on its interaction with myosin, and the result is a pulling force (Pollard, 1981).

Further elongation of the cell produced a concomitant stretching of the nucleus, and the perinuclear filaments congregated in the equatorial region to produce a contractile ring that constricted the nucleus (Figure 3c). Dumbbell configurations of nuclei analogous to this proto-telophase are characteristic of many lower eukaryotes with intranuclear mitosis (review: Heath, 1980), and although there is no evidence that it is the result of a constriction by a ring-shaped array of contractile filaments (e.g., Roos, 1975), such a mechanism is a serious possibility that may in fact have survived in extant organisms. Cytokinesis was in turn accomplished by a contractile ring constricting and cleaving the cell midway between the daughter nuclei (Figure 3d), as in modern animal

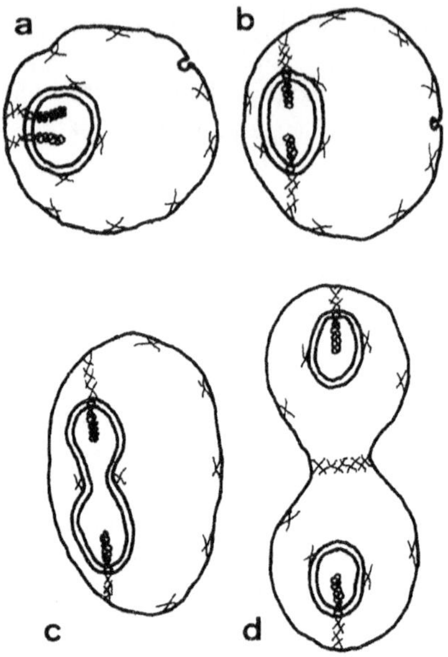

Fig. 3. Proto-mitosis. (a) The duplicated NAOs, to which the duplicated chromosomes are attached (for reasons of clarity only two are shown), move in opposite directions in the nuclear envelope. The system of contractile filaments provides the pulling force for this movement (cf. Fig. 2). (b) The NAOs have arrived at opposite 'poles' of the nucleus and the contractile system pulls it into an elongated shape while the cell also begins to elongate. (c) The cell elongates further and a perinuclear ring of contractile filaments constricts the nucleus approximately in the equatorial region. (d) Complete constriction of the nucleus has produced two daughter nuclei; a subplasmalemmal contractile ring now constricts and finally cleaves the cell into two daughter cells.

cells (Candle, 1980; Schroeder, 1975).

I think the model just described is attractive because the same motility system that is responsible for ameboid movement and cytosis also performs the intracellular movements required for nuclear division. The model is thus simpler than others proposing a combination of two motility systems, one based on contractile filaments for locomotion and cytosis, and another one based on MTs for genome segregation (Cavalier-Smith, 1975) or nuclear division (Margulis, 1970, 1981; Taylor, 1976). On the other hand, proto-mitosis was perhaps not a very reliable mechanism of genome segregation, for it lacked rigid axial elements that could impose a defined polarity on the movements of the daughter genomes.

5. Evolution of the Mitotic Spindle

I assume, as have others (Cavalier-Smith, 1975; Taylor, 1976), that MTs originated in the cytoplasm (Figure 4a). The primitive nucleus was certainly very much like the cytoplasm from which it was derived, and protein synthesis quite probably occurred in both cell compartments. If MTs evolved during this period, their appearance was

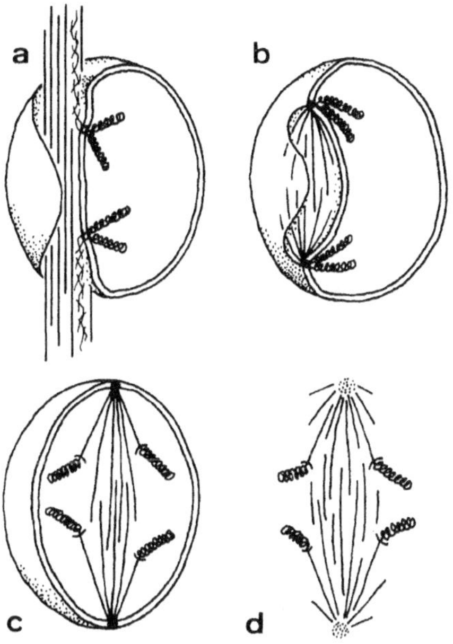

Fig. 4. Possible evolutionary sequence for the mitotic spindle (only the nucleus is shown). In (a-c) the right-hand portion of the nucleus is drawn as if sectioned at the level of the NAOs; for clarity only two pairs of chromosomes are shown inside the nucleus. (a) Cytoplasmic MTs form a bundle in a groove or cleft of the nucleus; the contractile filaments separate the NAOs together with the attached chromosomes along the axis the bundle defines. (b) A central spindle consisting of extranuclear MTs assists the segregation of the chromosomes that remain attached to the SPBs. (c) Engulfment of the spindle has placed it permanently inside the nucleus. The chromosomes have dissociated from the SPBs; kinetochores and kinetochore-MTs (kMTs) are new acquisitions. Chromosome segregation is by elongation of the central spindle alone, or by elongation in concert with poleward movement of the chromosomes. (d) The open spindle of higher eukaryotes, derived from the closed spindle of (c), or from the extranuclear spindle of (b) via a different evolutionary pathway (see text for additional explanations).

presumably as likely in the nucleus as in the cytoplasm, depending on the site of synthesis of their constituent proteins. However, I assume that MTs originated only after protein synthesis had become exclusively cytoplasmic and therefore consider it more plausible that assembly of MTs also evolved in the cytoplasm. Assembly inside the nucleus would have required the more complex processes of shuffling the RNA-messenger from the nucleus to the cytoplasm and the subsequent transport of the MT proteins into the nucleus. For intranuclear proteins such processes must have occurred, but in the case of MT proteins the additional difficulty would have been either to prevent their assembly until they were inside the nucleus, or to regulate assembly and disassembly in the cytoplasm *and* in the nucleus, tasks for which such a primitive cell may not have been equipped.

In the first step during the evolution of the mitotic spindle (Figure 4a), MTs may have played a static role (cf. Kubai, 1975), providing a rigid supporting system to ensure a more reliable equipartitioning of the chromosomes, which had increased in number as a consequence of the increasing complexity of the genome (Cavalier-Smith, 1975). In preparation for division, MTs aggregated to form a bundle near the nucleus, at the site of the NAO. The nuclear envelope invaginated to accommodate the NAO and the MT bundle in a groove or channel similar to the ones described in extant dinoflagellates (reviews: Kubai, 1975; Ris, 1975; Spector and Triemer, 1981). Sets of daughter chromosomes were collectively attached to the daughter-NAOs. Separation of the latter along the axis defined by the MT bundle, and concomitant chromosome segregation, occurred via specialized parts of the filament system connected to the NAOs (cf. Figure 3), but the defined polarity imposed by the MT bundle was a distinct advantage. At that point on the evolutionary pathway a truly eukaryotic system of contractile filaments similar to the actin-myosin system of extant eukaryotes had perhaps replaced the ancient primitive system of contractile filaments.

A second major step in the evolutionary sequence was possibly the formation of an extranuclear central spindle between the chromosome attachment points (Figure 4b). The spindle was located in a cleft of, or a channel in the nucleus, similar to the spindle of hypermastigote flagellates (but, of course, without the kinetochores and kinetochore MTs that these possess; Inoué and Ritter, 1978; Kubai, 1973). The two descendants of the NAO, the spindle pole bodies (SPBs), had acquired a new function, namely that of nucleating and organizing the spindle MTs, in other words they had differentiated into microtubule-organizing centers (MTOCs; Pickett-Heaps, 1969). The central spindle consisted of pole-to-pole MTs or of two interdigitating half-spindles. Segregation of the daughter chromosomes, still collectively attached to the SPBs, was initially accomplished by the concerted action of the contractile filament system and the central spindle. The spindle not only defined and maintained the axis and polarity of the division apparatus, but it also served to increase the distance between daughter genomes (cf. Figure 1c).

From this point on, divergent paths are imaginable, leading to spindles with quite different relative contributions of the filament and MT systems, but I assume the main trend was towards the development and functional integration of a chromosomal spindle, i.e. the acquisition of kinetochores and chromosomal MTs (Figure 4c, d). The intranuclear spindle (Figure 4c) is a possible intermediate on the pathway to the open spindle of higher eukaryotes (Figure 4d), but it could also have been a sideline of evolution. However, dissociation of the chromosomes from the SPBs was not without risk at this early stage, when the mechanism was not yet precisely tuned. It may thus have been safer to test and perfect it inside the nucleus. Various structural modifications of this spindle could have resulted in intranuclear spindles with intranuclear or extranuclear SPBs, etc. (cf. Heath, 1980). The translocation of the spindle into the nucleus can be considered the result of fusion of those parts of the nuclear envelope wrapping the spindle and dissolution of the engulfed parts of the membranes. The internalization, at each division, of the nascent spindle in some extant lower eukaryotes (see Heath, 1980, for examples) may be an ontogenetic vestige of these evolutionary events.

As the chromosomes separated from the SPBs, the capacity to function as MTOCs was also transmitted to them in the form of kinetochores (Figure 4c; Pickett-Heaps, 1974). The central spindle took over entirely the function of moving the SPBs apart. The chromosomal MTs (kMTs) initially played only the role of static anchoring structures (cf. Figure 1c), but the inevitable trend was towards a dynamic role for the kMTs, i.e. towards their participation in the poleward movement of the chromosomes (cf. Figure 1d).

6. Concluding Remarks

The origin and evolution of mitosis are inevitably speculative topics. Margulis (1981) and Goode (1975) have argued that mitosis is of polyphyletic origin. Considering the structural diversity of spindles in modern eukaryotes one is inclined to agree. However, as Oakley (1978) has rightly stated, and as Heath (1981a) has demonstrated, morphology of mitosis, when not related to function, is a phylogenetic criterion of limited value. As a criterion for evolutionary relationships it is hardly more valuable. I consider our present knowledge too restricted to rule out a monophyletic origin of mitosis, but I also accept Cavalier-Smith's (1978) objection to establishing "a linear progression of evolution based on modern mitotic variants".

The hypothesis I presented postulates the transition, concurrent with the evolution of the eukaryotic condition, from a mechanism of genome segregation mediated by the cell membrane to one operated by a system of contractile filaments. Like the models of Cavalier-Smith (1975, 1981) and Heath (1980), proto-mitosis (Figure 3) side-steps eukaryotic intermediates with a membrane-based mechanism of segregation (Kubai, 1975; Pickett-Heaps, 1975). It also offers an alternative to MTs as *the* supporting structures for the uncontested need of early eukaryotes to increase the distance between daughter genomes (Cavalier-Smith, 1975, 1978; Pickett-Heaps, 1974, 1975).

In agreement with Kubai (1975) I assume that MTs initially assisted karyokinesis as an extranuclear scaffold. Spindle evolution then possibly progressed from an extranuclear, non-chromosomal spindle to an intranuclear spindle including chromosomal MTs, and finally to the open spindles of higher eukaryotes. Pickett-Heaps (1974, 1975) has argued that closed spindles are more primitive than open spindles; he and other authors have also assumed an *intranuclear* origin for the MT spindle (Heath, 1980; Margulis, 1981). This concept contrasts with observations that many spindles considered primitive are *extranuclear* (Kubai, 1975; Ris, 1975).

Working out a hypothesis on the evolution of mitosis is a stimulating intellectual exercise, but, because rigorous testing is impossible, it is also frustrating. Heath's (1980) statement that "no fossil record is ever likely to yield useful information on ancestral mitotic systems" is certainly to the point. We must therefore look among present-day organisms for information concerning 'missing links'. Proto-mitosis (Figure 3) is a type of division not documented for any contemporary cell; unless it is found it remains purely hypothetical. The evolutionary stage depicted in Figure 4a, except for the collective attachment of the chromosomes, is similar to division in the dinoflagellate *Gyrodynium cohnii* (Kubai and Ris, 1969). The authors suggested that chromosome segregation in this possible intermediate between pro- and eukaryotes is mediated by the nuclear envelope; alternatively, a contractile system might be involved. Extranuclear spindles similar to the kind shown in Figure 4b exist in hypermastigote flagellates, but in these protists kinetochores occur in the nuclear envelope and kinetochore MTs play a role in chromosome segregation (Inoué and Ritter, 1978; Kubai, 1973). Given also the kinetochore-like structures and chromosomal MTs of many dinoflagellates (review: Spector and Triemer, 1981), this could mean that kinetochores evolved earlier than I assumed and that chromosomal MTs evolved outside, rather than inside the nucleus. In this case the intranuclear mitoses with spindle elongation and with or without poleward movement of chromosomes, as in many fungi and related organisms (review: Heath, 1981b), would be later evolutionary developments.

Future investigations will hopefully produce much solid data. Detailed studies within groups of organisms whose close taxonomic relationship is uncontested should assess the value of mitotic morphology for phylogeny and evolution and should include comparisons with more distantly related groups. Although perhaps not a very rewarding task, filling in the many gaps existing in the cytological record of protists relevant for phylogeny and evolution (cf. Heath, 1980, 1981a) could improve the situation considerably. Information on the division *mechanism* of dinoflagellates and spindle *function* in lower eukaryotes is highly desirable. Many organisms of interest are difficult objects for direct observations and experimentation, but, as the example of *Barbulanympha* shows (Inoué and Ritter, 1978), technical difficulties can be overcome. Last but not least, will the question of tubulin in prokaryotes soon be resolved, and will motility mechanisms involving non-actin and non-myosin filaments be biochemically characterized?

References

Allen, R. D. and Taylor, D. L.: 1975, *Molecules and Cell Movement*, Raven Press, New York, p. 239.
Allsop, A.: 1969, *New Phytol.* **68**, 591.
Amos, L.: 1975, *Molecules and Cell Movement*, Raven Press, New York, p. 411.
Cachon, J. and Cachon, M.: 1981, *BioSystems* **14**, 313.
Cachon, J., Cachon, M., Tilney, L. G., and Tilney, M. S.: 1977, *J. Cell Biol.* **72**, 314.
Cande, W. Z.: 1980, *J. Cell Biol.* **87**, 326.
Cavalier-Smith, T.: 1975, *Nature* **256**, 463.
Cavalier-Smith, T.: 1978, *BioSystems* **10**, 93.
Cavalier-Smith, T.: 1981, *Soc. Gen. Microbiol. Symp.* **32**, 33.
Fracek, S. P., Jr.: 1981, *Abstr. EMBO Workshop Microtubules in Microorganisms*, 2.
Gilbert, M. and Fulton, A. B.: 1982, *J. Cell Biol.* **95**, 283a.
Girbardt, M. and Hädrich, H.: 1975, *Z. allg. Mikrobiol.* **15**, 157.
Goode, D.: 1975, *BioSystems* **7**, 318.
Gray, M. W. and Doolittle, W. F.: 1982, *Microbiol. Rev.* **46**, 1.
Heath, I. B.: 1980, *Int. Rev. Cytol.* **64**, 1.
Heath, I. B.: 1981a, *BioSystems* **14**, 261.
Heath, I. B.: 1981b, *The Fungal Nucleus*, Cambridge University Press, Cambridge, p. 85.
Hobbs, V. S., Jenkins, R. A., and Bamburg, J. R.: 1983, *J. Cell. Sci.* **60**, 169.
Inoué, S. and Ritter, H., Jr.: 1978, *J. Cell Biol.* **77**, 655.
Jacob, F., Brenner, S., and Cuzin, F.: 1963, *Cold Spring Harb. Symp. Quant. Biol.* **28**, 329.

Kubai, D. F.: 1973, *J. Cell Sci.* **13**, 511.
Kubai, D. F.: 1975, *Int. Rev. Cytol.* **43**, 167.
Kubai, D. F. and Ris, H.: 1969, *J. Cell Biol.* **40**, 508.
Maniloff, J.: 1981, *BioSystems* **14**, 305.
Margulis, L.: 1970, *Origin of Eukaryotic Cells*, Yale Univ. Press, New Haven.
Margulis, L.: 1981, *Symbiosis in Cell Evolution*, Freeman, San Francisco.
Margulis, L., To, L., and Chase, D.: 1978, *Science* **200**, 1118.
Margulis, L., To, L., and Chase, D.: 1981, *Ann. New York Acad. Sci.* **361**, 356.
Nelson, G. A., Roberts, T. M., and Ward, S.: 1982, *J. Cell Biol.* **92**, 121.
Oakley, B. R.: 1978, *BioSystems* **10**, 59.
Pickett-Heaps, J. D.: 1969, *Cytobios* **1** (3), 257.
Pickett-Heaps, J. D.: 1974, *BioSystems* **6**, 37.
Pickett-Heaps, J. D.: 1975, *Ann. New York Acad. Sci.* **253**, 352.
Pollard, T. D.: 1981, *J. Cell Biol.* **91** (3, pt. 2), 156s.
Ris, H.: 1975, *BioSystems* **7**, 298.
Roos, U.-P.: 1975, *J. Cell Biol.* **64**, 480.
Routledge, L. M.: 1978, *J. Cell Biol.* **77**, 358.
Ryter, A.: 1968, *Bacteriol. Rev.* **32**, 39.
Schroeder, T. E.: 1975, *Molecules and Cell Movement*, Raven Press, New York, p. 305.
Schwemmler, W.: 1979, *Mechanismen der Zellevolution*, Walter de Gruyter, Berlin.
Searcy, D. G., Stein, D. B., and Searcy, K. B.: 1981, *Ann. New York Acad. Sci.* **361**, 31.
Spector, D. L. and Triemer, R. E.: 1981, *BioSystems* **14**, 289.
Taylor, F. J. R.: 1974, *Taxon* **23**, 229.
Taylor, F. J. R.: 1976, *Taxon* **25**, 377.
Tilney, L. G.: 1978, *J. Cell Biol.* **77**, 551.
Uzzell, T. and Spolsky, C.: 1974, *Amer. Sci.* **62**, 334.

THE FIFTH BASE: A NATURAL FEATURE OF DINOFLAGELLATE DNA

COLETTE GALLERON

Laboratoire de Biologie Cellulaire Végétale, ERA 325 du CNRS, Université de Paris-VII, 2 Place Jussieu, (Tour 53–54, 3° Et.) 75251 Paris Cedex 05, France

(Received 12 December, 1983)

Abstract. The presence of 5-Hydroxymethyluracil which replaces important fractions of thymine in Dinoflagellate DNA is analyzed according to different procaryotic models of biosynthesis. The detection of HOMeU in *Amphidinium carterae* cells suggests an incorporation of this 5th base during polymerisation of DNA chains rather than a post-synthetic modification of thymines. A relationship between HOMeU and the permanent repair like DNA synthesis observed in *A. carterae* is discussed.

1. Introduction

Modified bases are a common feature in DNA. It is also well-known that except in some cases, atypical bases are detected in small amounts. They can enter in the DNA by three pathways (Kornberg, 1980): (1) incorporation of a synthetic base, nucleoside or nucleotide analog; (2) incorporation of an enzymatically altered nucleotide; (3) post-synthetic modification of bases in DNA chains.

In the first case, the base analog is converted by a salvage pathway to a nucleotide 5′-tri-phosphate which can effectively compete with the natural nucleotide for base-pairing. This process has been largely used for experimental studies of DNA synthesis by incorporation of labelled or heavy analogs, in procaryotes and eucaryotes. In three species of dinoflagellates, DNA synthesis has been explored by this method (Franker *et al.*, 1974; Galleron *et al.*, 1975; Filfilan and Sigee, 1977; Galleron and Durrand, 1978, 1979).

The best examples of the second kind are certain phage DNAs which include uncommon bases (Uracil and 5-HOMeUracil in some *B. subtilis* phages, 5-HOMeCytosine in T-even phages of *E. coli* and in certain phages of *X. orizae*) even to the exclusion of thymine or cytosine. The biosynthetic pathways of HOMeC, HOMeU and U are well known and all the corresponding enzymes have been characterized (Mathews *et al.*, 1964; Reeve *et al.*, 1978). In the case of SPO1 and other closely-related phases of *B. subtilis*, the genes controlling DNA synthesis have been located in the genome. Although the pyrimidine pathway is altered in 5 steps to substitute 5-HOMeU to T, inclusion of thymine residues is tolerated up to 20% of HOMeU a proportion close to that observed in dinoflagellates. In SPO1, a phage-encoded DNA-polymerase incorporates dHOMeUTTP, dTTP, dUTP, and 5-BrUdR at identical rates (Yehle and Ganesan, 1973).

In bacterial cells as well as in animal cells, the relative pool sizes of dUTP and dTTP regulate the incorporation of uracil into DNA. The level of two different enzymes helps

Origins of Life **13** (1984) 195-203. 0302-1688/84/0133-0195$01.35.

to eliminate the foreign base: when the thymidilate synthetase pathway is inhibited by methotrexate, the ratio of dUTP/dTTP in the cellular pool is raised up to 20% in spite of dUTPase which normally degrades almost completely dUTP (Nilsson *et al.*, 1980). However, under normal circumstances, it is the level of uracil-N-glycosylase, a non site-specific excision enzyme which determines the persistence of uracil in DNA. Bacterial mutants defective in this enzyme as well as methotrexate treated-cells can accumulate uracil to levels comparable with thymine levels. No other glycosylase removing pyrimidines has been reported in the literature while uracil-N-glycosylase has been found in some bacteria, phages and mammals (Kornberg, 1980).

Post-modification of the DNA occurs through the action of site-specific enzymes which protect phage and bacterial DNAs against restriction by host and phage-encoded nuclease. In phage T_4, for instance, α- and β-glycosyl-transferases (Josse and Kornberg, 1962) coded by the phage DNA which glycosylates the HOMeC residues are among the most important examples together with type I and type II restriction-modification methylases of *E. coli*.

Rae, in 1973, discovered that in the DNA of a marine dinoflagellate, *Crypthecodinium cohnii*, as much as 37% of thymine is replaced by HOMeU. High rates (up to 68% in *Prorocentrum micans*) have been measured since (Rae, 1976; Galleron and Durrand, 1978; Herzog and Soyer, 1982). Looking for a biological function of HOMeU by analyzing *C. cohnii* DNA, Rae and Steele (1978), found no involvement of this base in the nucleoid-like structure of the chromosomes, and no true restriction activity, i.e. no set of conditions in which nuclease activity of *C. cohnii* extracts degraded preferentially heterotypic DNA. However, he did find a definite non-randomness of the positions of HOMeU in DNA sequences. These observations led him to make the following hypothesis: the uncommon base might be the last remnant of an ancient restriction-modification system once possessed by proto-dinoflagellates.

Looking for the presence of HOMeU while exploring the cycle of DNA synthesis in *Amphidinium carterae* synchronized cultures, we made an assumption opposite to Rae's hypothesis, and tried to verify whether the large fraction (HOMeU/HOMeU + T = 62%) of HOMeU found in *A. carterae* DNA were incorporated through a phage-like mechanism.

2. Methods

2.1 Cultures of Amphidinium carterae (Hulburt) and extraction of DNA

Synchronized cultures, DNA extraction and purification, isopycnic centrifugation in CsCl gradients, *in vivo* labelling of cells by pyrimidine precursors or analogs and radioactive counting of liquid fractions have been performed as previously described (Galleron and Durrand, 1978, 1979).

2.2. Preparation of the acid-soluble pool of nucleotides

It was made by dissolution in 10% TCA of a cell pellet, radioactively labelled *in vivo* and extensively washed in culture medium until no radioactivity appeared in discarded liquids. Cells were ground on ice and depigmented in methanol, ethanol and ether. Nucleotides (1-, 2- and 3-phosphates) were separated from the TCA-soluble fraction by ion-exchange chromatography: 100 µl-samples were put on top of small columns (Pasteur pipettes) loaded with Dowex 50 AG 500 WX4, Hydrogen form (Bio-Rad) in 0.1 M formic acid. Nucleosides were eluted first, then nucleotides with 0.1 M HCl. 95% of the radioactivity was recovered in the 3 peaks.

2.3. Acid hydrolysis of DNA

Acid hydrolysis of DNA (extensively dialyzed against water after CsCl centrifugation) and of nucleotides extracted from the cellular pool were performed on lyophylized samples. They were dissolved in 0.5 ml 1 M HCl and boiled in sealed glass capsules. The hydrolysis was stopped at −20 °C in ethanol and dry ice.

2.4. Enzymatic hydrolysis of DNA

In vivo labelled DNA was enzymatically hydrolyzed with DNase I and snake venom phosphodiesterase according to Ray and Hanawalt (1964). Nucleotides obtained were treated with alkaline phosphatase (Sigman, 50 µg/ml, 2 h incubation at pH 10, 37 °C). This treatment could not be applied to TCA-soluble fractions for which pH adjustment led to an excessive dilution of samples.

2.5. Radioactive precursors

^{3}H-Methyl-Thymidine (^{3}H-Me-TdR), (^{3}H-6-deoxyUridine (^{3}H-6-UdR), ^{3}H-6-Uracil (^{3}H-6-U) and ^{14}C-5-HydroxyMethylUracil (^{14}C-5-HOMeU) were purchased from CEA (Saclay, France); ^{3}H-6-Bromo-5-deoxyUridine (^{3}H-Br-5-UdR) was purchased from Amersham (France). Radioactive molecules were given to cultures under conditions defined in Table I.

2.6. Base and nucleotide separations

They were made by thin layer chromatography (Randerath, 1971). Samples, filtered (bases) or acid-eluted, (nucleosides) from Dowex columns were evaporated to concentrations of 10 or 20 µg/100 µl of 0.1 M HCl. 5 µl spots were chromatographed in 2 dimensions on Kieselgel plates (60 F 254, Merck). The solvents were: first dimension; chloroform-methanol-H_2O (4 : 2 : 1); second dimension; ethyl acetate-isopropanol-H_2O (75-18-9). Spots were identified by co-migration with the corresponding commercial product visualized under UV long-wave illumination. Bases and nucleosides (T, C, U) were purchased from Sigma. HOMeU and HOMeU-deoxyUridine were purchased from Serva (Tebu-France). The HOMeU nucleotide was not commercially available.

2.7. Control RNA extractions

They were made according to Laulhère and Rozier (1976). Radioactive counting of solid samples was performed by elution of the spots in 10 ml Soluene (Packard) and counting in a Packard model 2425 Tricarb.

3. Abbreviations

^3H: tritium; ^{14}C: radioactive carbon; C: cytosine; H, HOMeU, 5-HOMeU: 5-Hydroxymethyluracil; T: thymine; TdR: thymidine; U: uracil; UdR: deoxyuridine; Cpm: counts per minute; TCA: trichloracetic acid.

4. Results

4.1. In vivo incorporation of labelled analogs

We had previously observed that *A. carterae* cells incorporate very easily all precursors of the pyrimidine biosynthesis pathway in their DNA (Galleron and Durrand, 1978). We knew also that a permanent incorporation is superimposed to the S-phase, a

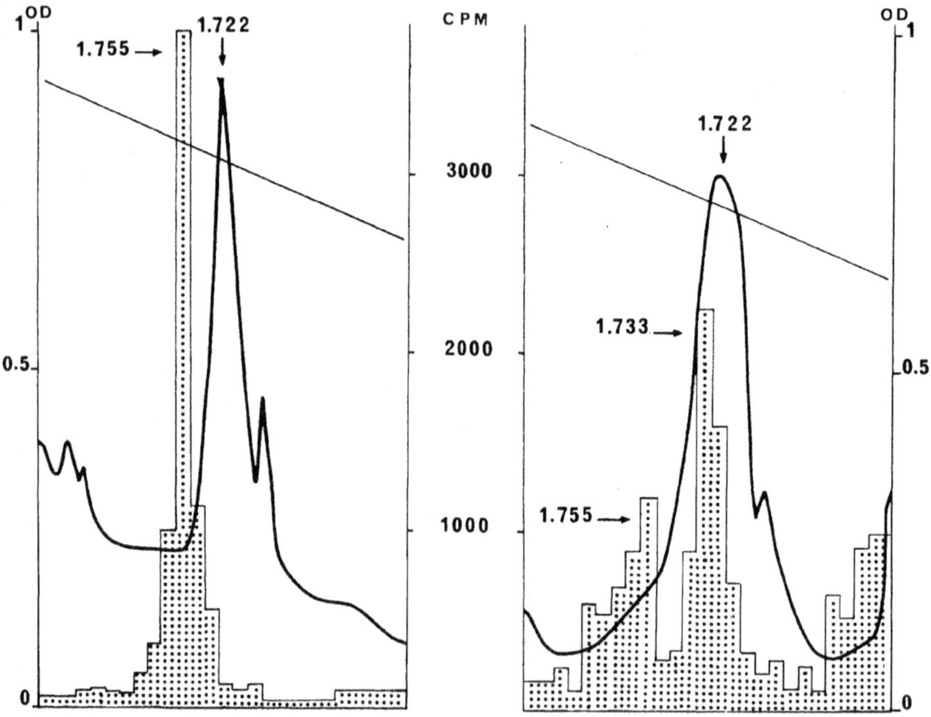

Fig. 1. CsCl centrifugation profile of *Amphidinium carterae* DNA, labelled in vivo with ^{3}H-6-Br-5-UdR. Four synchronized 1.5-1 cultures of *A. carterae* were pooled and incubated for 4 h with ^{3}H-6-Br-5-UdR; left: cells in S-phase; right: cells in G_2 phase. Nuclear DNA was isolated by centrifugation in CsCl then recentrifuged with cold *A. carterae* nuclear DNA as a density marker (d = 1.722, Galleron and Durrand, 1978). 20 drop-fractions were collected and counted for radioactivity.

[dotted box] : histogram of Cpm;
——— : O.D. at 260 nm.

phenomenon detectable in well-synchronized cultures (Filfilan and Sigee, 1977; Galleron and Durrand, 1979). Figure 1 shows a centrifugation profile in CsCl of *A. carterae* DNA labelled *in vivo* with tritiated-bromo-deoxyuridine. The DNA extracted from cells in S-phase is heavier than the normal nuclear *A. carterae* DNA, as expected. For cells in G_2 phase, two DNA peaks are observed, one of which, the heavier, is probably due to a slightly imperfect synchronization of cells; the other one has a density close to, but different from the density of *A. carterae* nuclear DNA (d = 1.722, Galleron and Durrand, 1978). This slightly heavier peak corresponds to an unscheduled DNA synthesis insensitive to ethidium bromide inhibition (Galleron and Durrand, 1979) and appears to be a large scale-repair mechanism.

TABLE I

In vivo labelling of *Amphidinium carterae* DNA cultures with pyrimidine precursors and analogs. In all experiments, DNA was extracted, CsCl centrifuged and counted for radioactivity as cited in 'Methods'. Counts per minute (Cpm) are given after subtraction of background (less than 25 Cpm under our conditions).

	Precursor or Thymidine analog	Spec. Act. μCi/ml	Incubation time	Chase	Spec. Act. of DNA Cpm/μg
I	^{3}H-6-Uracil (43 Ci/mM)	2.4	10 min.	1 h	10^7
II	^{3}H-Me-Thymidine (43 Ci/mM)	0.6	3h	wash	4×10^5
III	^{3}H-6-deoxyUridine (23 Ci/mM)	0.6	3h	wash	5×10^4
IV	^{14}C-5-Hydroxy-Me-Uracil (25 Ci/mM)	0.1 0.6	4h 6h	wash wash	background 500
V	^{3}H-6-Br-5-deoxy-Uridine (21 Ci/mM)	0.625	4h	wash	5000 (S-phase) 2000 (G_2-phase)

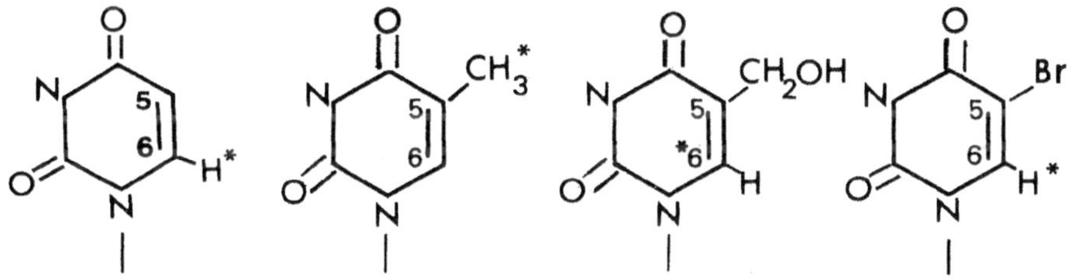

I, III II IV V

4.2. BASE AND NUCLEOSIDE ANALYSIS IN THE CELLULAR POOL OF NUCLEOTIDES

Table I shows the efficiency with which *A. carterae* DNA is labelled by bases, nucleosides and analogs. Figure 2, Figure 3 and Table III show the labelling of bases and nucleosides of the DNA and the labelling of the bases hydrolyzed from the pool of nucleotides (see 'Methods') in two different experiments.

Table III A displays the labelling of pyrimidines after a 2 h incubation of *A. carterae* cultures during S-phase. Thymine and HOMeU are labelled by ^{3}H-6-UdR and also by ^{3}H-Me-TdR with a higher rate of synthesis of HOMeU in the pool and of T in the DNA (unlike the relative contents of T and HOMeU in *A. carterae* DNA) HOMeU cannot be labelled by ^{3}H-Me-TdR through a degradation of T and a new utilization of the free methyl group which otherwise would label DNA. Control RNA extractions have been entirely negative concerning this point.

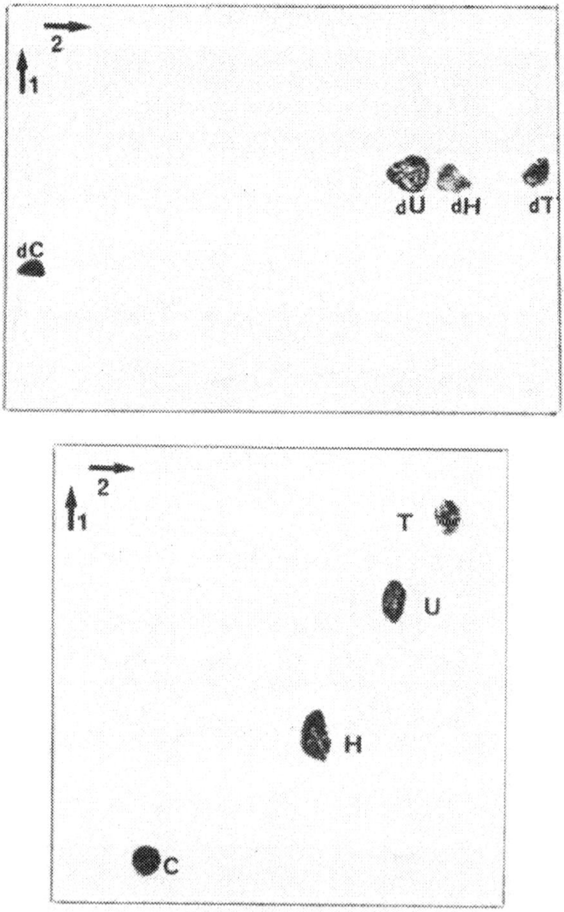

96

Fig. 2. Separation of pyrimidine nucleosides *(A)* and pyrimidines (B) by thin layer chromatography. Migration in two dimensions and visualization of spots have been made as described in 'Methods'. *Top*: separation of nucleosides; *bottom*: separation of bases.

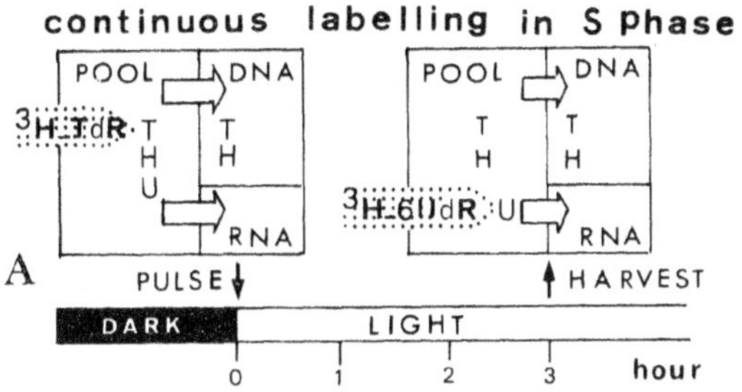

TABLE IIIA

Labelling	Hydrolysate		Thy	HOMeU	Cyt	Ura
^{3}H-Me-TdR	DNA	Nucleosides	1000	100	0	
^{3}H-Me-TdR		Bases	550	0	0	
	Pool	Bases	740	960	0	0
^{3}H-6-UdR	DNA	Nucleosides	730	510	0	
^{3}H-6-UdR		Bases	530	300	90	
	Pool	Bases	540	5300	140	20

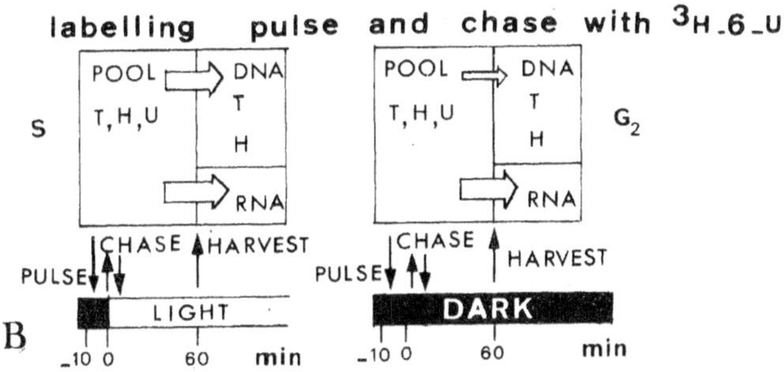

TABLE IIIB

Cell cycle Phase	Hydrolysate		Thy	HOMeU	Cyt	Ura
	DNA	Nucleosides	10900	7000	1600	
S		Basis	3100	1500	3000	
	Pool	Bases	100	24000	1300	1000
	DNA	Nucleosides	670	850	25	
G_2		Bases	950	1500	250	
	Pool	Bases	50	650	0	100

Fig. 3. and Table III. Radioactivity of pyrimidine nucleosides (Cpm) measured in hydrolysates of DNA and nucleotide pool separated by thin layer chromatography.
(A) Three 1l-cultures were labelled *in vivo* with: *left*: ^{3}H-Me-TdR and *right*: ^{3}H-6-UdR for 3 h under conditions described in Table I and according to the schedule drawn on the Figure above.
(B) In three 1l-cultures, 500 ml-samples were labelled with ^{3}H-6-U (see Table I) and successively harvested in S- and G_2-phases. After 10 min-pulses, ^{3}H-6-U was chased with cold uracil for 1 h. The DNA and nucleotide pool was thus extracted, hydrolyzed and analyzed as described above. Spots on chromatograms were excised and counted for radioactivity as described in 'Methods'.

As it can be seen in Table IIIB, a strong labelling of pyrimidines has been detected one hour after a 10 min-pulse of ^{3}H-6-U (followed by a chase). The proportion of the labels of T and HOMeU in the pool is still opposite to the proportion measured in the DNA. HOMeU is highly labelled and accumulates in the pool. During G_2-phase, the reduced synthesis does not lead to a labelling of thymine.

Considered together, these observations suggest an important synthesis of HOMeU through an enzymatic pathway in the cellular pool, and a degradation of thymine residues transformed into HOMeU as is shown by the labelling of HOMeU by ^{3}H-Me-TdR.

5. Discussion

Even in well-growing cultures of *A. carterae*, the mixture of mono-, di- and tri-phosphates that we extract from the acid-soluble fraction was not enough to allow the observation of isolated tri-phosphates on a chromatogram, after hydrolysis to bases. Nevertheless, HOMeU which is so highly labelled by exogenous thymidine, deoxy-uridine, or uracil is very abundant in the cellular pool. The fact that the labelling of an exogenous HOMeU given to cultures is not observable in the DNA could be explained by the dilution of this base in the cold cellular pool.

In *A. carterae*, *C. cohnii* and *P. micans* which incorporate so efficiently pyrimidine precursors (Franker *et al.*, 1974; Galleron *et al.*, 1975; Filfilan and Sigee, 1977), there is nothing in our knowledge that could prevent algal DNA-polymerases from taking up HOMeU nucleotides as building blocks for the DNA. On the other hand, the finding by Rae (1978) of non-random positions of HOMeU in *C. cohnii* DNA, must be considered by reference to the natural DNA template. Sequencing large repetitive fractions of Dinoflagellate nuclear genome is not feasible (*A. carterae* DNA: 1.9×10^{12} daltons per haploid cell, *C. cohnii*: 4.2×10^{12}; *P. micans*: 2.5×10^{12}, Loeblich, 1976). All we know about sequence organization is the preferential digestion of *C. cohnii* and *P. micans* total nuclear DNA by restriction enzymes which contain no T-T sequences in their total recognition site (Steele, 1980, Galleron, unpublished results). These observations are coherent with the finding by Rae (1978) of a very low frequency of 5′-HOMeU-p-T-3′ in *C. cohnii* DNA (7.9% instead of the 29.3% expected from GC content if HOMeU was randomly located in the genome).

In the absence of evidence for any other biological function of HOMeU in dinoflagellate DNA, we are still left with Rae's hypothesis: the persistence of a protection device against foreign endonucleases. Among the five dinoflagellate species in which HOMeU has been detected only one (*Symbiodinium microadriaticum*, Rae, 1976) is parasitic. Therefore, we cannot eliminate the interpretation of HOMeU in DNA as a vestige left in these organisms which once lived in associations either as hosts, symbionts or parasites (Sarjeant, 1974; Loeblich, 1976) and were thereby exposed to foreign endonucleases.

Further work should decide whether there is a competition between T and HOMeU regulated by their relative pool sizes, as in phage SPO1. It could be verified if, in methotrexate treated-cells, the levels of HOMeU in DNA were raised by an inhibition of the thymidilate synthetase pathway. Southern hybridization between digested total DNA and probes made of cloned parts of the SPO1 genome coding for the enzymes of HOMeU synthesis could also bring valuable information.

Moreover, if the incorporation of pyrimidine precursors throughout the cell cycle corresponds to a repair mechanisms after preferential excision of thymines from DNA, it must work through a site-specific endonuclease still undiscovered, or through a base-specific enzyme similar to the ubiquitous uracil-N-glycolase.

Answers to these questions should give some clues to solve the mystery of the presence of HOMeU in dinoflagellate DNA, one of the strangest features of these primitive eukaryotes.

Acknowledgments

The author wishes to thank C. Guennou for helpful technical assistance and Drs P. M. M. Rae, D. Sigee and M. Herzog for fruitful discussions.

References

Filfilan, S. A. and Sigee, D. C.: 1977, *J. Cell Sci.* **27**, 81.
Franker, C. K., Sakharani, M. L., Prichard, C. D., and Lamden, C. A.: 1974, *J. Phycol.* 7, 165.
Galleron, C. and Durrand, A, M.: 1978, *Biochimie* **60**, 1235.
Galleron, C. and Durrand, A. M.: 1979, *Protoplasma* **100**, 145.
Galleron, C., Kovoor, A., and Durrand, A. M.: 1975, *XIIth Int. Botanical Cong.* Leningrad, n° 56.
Herzog, M. and Soyer, M. O.: 1982, *Eur. J. Cell Biol.* **27**, 151.
Josse, J. and Kornberg, A.: 1962, *J. Biol. Chem.* **237**, 1968.
Kornberg, A.: 1980, *DNA Replication*, Freeman and Co, San Francisco, p. 649.
Loeblich, III, A. R.: 1976, *J. Protozool.* **23**, 13.
Mathews, C. K., Brown, F., and Cohen, S.S.: 1964, *J. Biol. Chem.* **239**, 2957.
Nilsson, S., Reichard, P., and Skoog, L.: 1980, *J. Biol. Chem.* **255**, 9552.
Rae, P. M. M.: 1973, *Proc. Nat. Acad. Sci. U.S.A.* **70**, 1141.
Rae, P. M. M.: 1976, *Science* **194**, 1062.
Rae, P. M. M. and Steele, R. E.: 1978, *Biosystems* **10**, 37.
Randerath, K.: 1971, *Chromatographie sur Couches Minces*, Gauthier-Villars, Paris, p. 221.
Ray, D. S. and Hanawalt, P. C.: 1964, *J. Mol. Biol.* **9**, 812.
Reeve, J. N., Mertens, G., and Amann, E.: 1978, *J. Mol. Biol.* **120**, 183.
Sarjeant, W. A.: 1974, *Fossil and Living Dinoflagellates*, Academic Press, London, p. 118.
Steele, R. E.: 1980, *Thesis*, New Haven, unpublished, p. 150.
Yehle, C. O. and Ganesan, A. T.: 1973, *J. Biol. Chem.* **248**, 7456.

ULTRASTRUCTURAL AND BIOCHEMICAL NUCLEAR ASPECTS OF EUKARYOTE CLASSIFICATION: INDEPENDENT EVOLUTION OF THE DINOFLAGELLATES AS A SISTER GROUP OF THE ACTUAL EUKARYOTES?

MICHEL HERZOG, SIGURD VON BOLETZKY and MARIE-ODILE SOYER
C.N.R.S., LA 117, Université Pierre et Marie Curie, Laboratoire Arago, F-66650 Banyuls-sur-Mer, France

(Received 12 December, 1983)

Abstract. Structural and functional features of the dinoflagellate nucleus are examined and compared to those commonly found in Prokaryotes and in Eukaryotes. It appears that dinoflagellate protists, while showing several ancestral characters also found in Prokaryotes are above the prokaryote level in terms of their organization but below that of the other Eukaryotes. Some characters are typical of dinoflagellate nuclei alone, and no correspondence is found in either bacterial nucleoïd or typical eukaryote nuclei. This supports Loeblich's (1976) proposal that dinoflagellate evolution may have been independent of that of the Eukaryotes. This concept can now be refined using an argumentation plan *sensu* Hennig (Hennig and Schlee, 1978) and appears to be in accordance with the Mesokaryote model introduced by Dodge (1965).

1. Introduction

Dinoflagellate protists are a large, highly diversified group of planktonic algae which comprise a variety of morphologies and life styles, and different levels of complexity. Paleontological investigations have shown that Dinoflagellates form an important phytoplankton group since the Jurassic (Sarjeant, 1974), but some primitive aspects, in particular of their nuclear structure and composition, suggest a considerably earlier origin. This is especially clear in the *Prorocentrales*, which are considered to be the most primitive living order within the Dinoflagellates (Loeblich, 1976; Taylor, 1980).

As early as 1920, Chatton considered the dinoflagellate nucleus as unique among cell nuclei, and proposed the term 'dinokaryon'. Chatton (1937) also made the distinction between 'prokaryotic' and 'eukaryotic protists'; Dinoflagellates are characterized by a well-defined nucleus and by this sign belong to the latter category. The first detailed ultrastructural observations of the dinokaryon were published by Grassé and Dragesco (1957), and by Grell and Wohlfarth-Bottermann (1957); these revealed the characteristic arch-shape aspect of their chromosomes as observed only in thin sections. Cytochemical data from Ris (1962) and Dodge (1964) pointed out another peculiar feature of these permanently condensed chromosomes: the absence of cytochemically detectable basic proteins (histones), except for the parasitic species *Syndinium* sp. (Ris and Kubai, 1974). Histones are common constituents of eukaryotic chromatin and form the protein part of the nucleosomes (for a review see Kornberg, 1977).

It is generally accepted that the dinoflagellate nucleus is above prokaryote level in terms of organization, but below the level of the other Eukaryotes. Questions

Origins of Life **13** (1984) 205-215. 0302-1688/84/0133-0205$01.65.

remaining at present thus concern the position of this group among the 'lower Eukaryotes'. Taylor (1978) considered Dinoflagellates as the most primitive eukaryotic cells, but viewing all their cellular features has not so far resulted in a clear picture of their position.

Our approach to this problem will be to concentrate on the nuclear features as a guideline for further discussions, essentially in the way Dodge and Loeblich have outlined the problem. The concept of 'Mesokaryotes' based essentially on their available ultrastructural and cytochemical data (Dodge, 1965) has been refined by Loeblich (1976) on biochemical grounds. Loeblich suggested that Dinoflagellates "are a geologically old group, one that perhaps diverged from the higher eukaryotic lineage before evolution of eukaryotic chromatin but after the evolution of repeated DNA". This phylogenetic model has been called in question by Cavalier-Smith (1981) with the argument that the lack of histones (and nucleosomes) in dinoflagellate chromatin could be explained by deletion of the clustered histone genes typical of all other Eukaryotes. This hypothesis is put forward quite categorically, but as far as we can see, without objective support.

The aim of our contribution is to add a few new aspects to the Mesokaryote concept by deliberately focussing our attention on the typical nuclear features of Dinoflagellates as far as they can be now defined in macromolecular, fine structural and functional terms. We then compare these features with typical nuclear conditions of both Prokaryotes and Eukaryotes. From this basically typological approach, we attempt to proceed towards a phylogenetically consistent picture by sorting the nuclear features according to aspects of plesiomorphy (common ancestral features) and apomorphy (uniquely derived features) following the views of Hennig and Schlee, 1978. This attempt is made in full awareness of the problems that will have to be solved later by a detailed analysis of structural and functional correlations between nuclear and extranuclear features. But we think that the analysis has to set out from the central apparatus of the cell, i.e. the nucleus as the carrier of genetic information.

We tentatively suggest that there is a monophyletic group of unicellular organisms comprising most of the actually free-living Dinoflagellates, and that it forms a sister group to the 'higher' Eukaryotes. We feel that this hypothesis, which is basically identical to the mesokaryote model, is of greater promise for further clarifications than evasive arguments about overall similarities.

2. Eukaryotic Features Observed in Dinoflagellate Nuclei

Several nuclear features of the Dinoflagellates are very similar to those of the other Eukaryotes, as suggested by numerous ultrastructural studies published over the last forty years. Dinoflagellates have a membrane-bounded nucleus (Grell and Schwalbach, 1965), with highly organized chromosomes (Figure 1). They also have one or several nucleolar structures (Figure 1b) generally in close contact with chromosomes (Soyer and Haapala, 1974). Although a mitotic spindle does not exist, long extra-nuclear microtubules could act as an extra-nuclear spindle; this is at least conceivable for a few species (Oakley and Dodge, 1979). In other species, however, a more primitive mechanism seems to exist (see below). More recent molecular data again indicate some eukaryotic features in the dinokaryon. Firstly, a distinct S-phase of DNA synthesis has been demonstrated in several partially synchronized species (Franker, 1971; Franker *et al.*, 1974; Galleron and Durrand, 1979). Secondly, renaturation studies published by several authors (Britten and Kohne, 1968; Allen *et al.*, 1975; Steele, 1980; Hinnebusch *et al.*, 1980) demonstrate the presence of a large amount of repeated DNA (50 to 60 %) interspersed in a manner similar to the other Eukaryotes, with a highly complex non-repetitive DNA. Finally, Steele (1980) using recombinant DNA techniques, reported that in *Crypthecodinium cohnii* ribosomal DNA (rDNA) exists as a complex gene family organized in tandemly repeated 15 Kb units, as in other eukaryotic rDNA.

One could also take ribosomal RNA into consideration for which molecular weight, sedimentation coefficient (S), base methylation and sequence homologies suggest that dinoflagellate rRNA represents an early form of eukaryotic rRNA (Rae, 1970; Gressel *et al.*, 1975; Werner-Schlenzka *et al.*, 1978; Hinnebusch *et al.*, 1981). Also, Reddy *et al.* (1983) demonstrated the presence of six capped small nuclear RNAs in *C. cohnii* similar in several respects to the U_1 to U_6 Sn RNAs of higher Eukaryotes.

3. Prokaryotic Features Observed in Dinoflagellate Nuclei

Along with the eukaryotic features, dinoflagellate nuclei display several more primitive traits, in particular the arch-shaped fibrillar appearance (Figure 1a, b) of the thin sectioned chromosomes which are reminiscent of the bacterial nucleoïd (Giesbrecht, 1965; Gourret, 1978). For some species it has been shown that chromosome attachment on the nuclear membrane is concomitant with the chromatid segregation. In functional terms, this could be compared with the corresponding mechanism in bacteria where the dividing nucleoïd is attached to the plasmic membrane.

Moreover, we have recently been able to show that the stabilization of the chromosomal fabric is maintained by divalent cations (Herzog and Soyer, 1983) and by structural RNA molecules (in prep.). Involvement of the latter was demonstrated also in the stabilization of the supercoiling of DNA loops in bacteria (Pettijohn *et al.*, 1973).

Nucleofilament structure also displays prokaryotic characters as demonstrated by chromatin spreading and biochemical investigations. Chromatin is composed of smooth filaments, 3–6 nm in diameter, which do not present the typical eukaryotic 'beads-on-a-string' appearance (Hamkalo and Rattner, 1977; Rizzo and Burghart, 1980, 1982; Herzog and Soyer, 1981). Absence of nucleosomes was confirmed (Bodansky *et al.*, 1979; Rizzo, 1981; Herzog and Soyer, 1981) by demonstrating the lack of histones (basic nuclear proteins of Eukaryotes) and the absence in nuclease-treated nuclei of the DNA fragment repeats which in true Eukaryotes are protected by the nucleosome structure. On the other hand, low amounts of basic nuclear proteins (12 000 and 13 000 daltons) were detected in several dinoflagellate species. In the case of *Crypthecodinium cohnii*, Rizzo and Nooden (1974) indicated that their composition in amino acids was different from that of histones. The possible relationship with HU

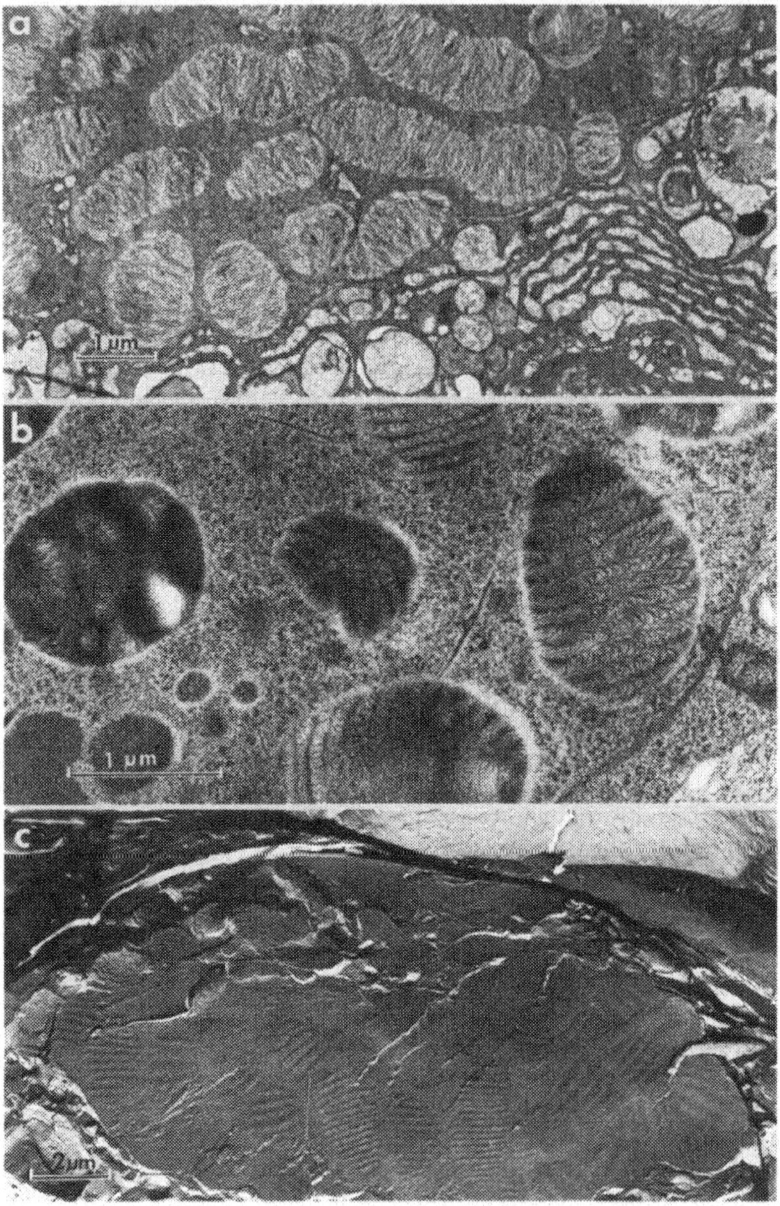

Fig. 1. The typical arch-shape appearance of *Prorocentrum micans* chromosomes when thin sectioned (a) after chemical fixation (Soyer, 1977) or (b) after cryofixation (Escaig *et al.*, 1977). In (c), a freeze-fractured nucleus in which chromosomes show transverse striation with a periodicity of about 150 to 200 nm. (a) 11 000 × ; (b) 21 000 × ; (c) 5 400 × .

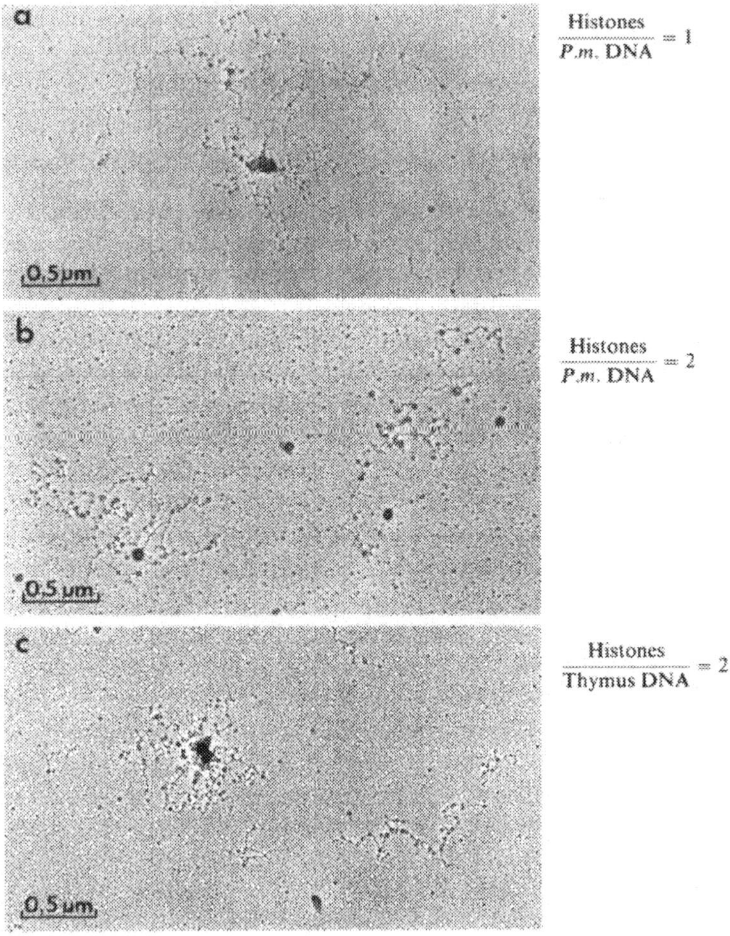

Fig. 2. *In vitro* reconstitution of nucleosomes in heterologous conditions using a mixture of purified corn histones, without histone H1 (a gift from Dr C. Gigot), and sonicated DNA from (a – b) the Dinoflagellate *Prorocentrum micans* or (c) from calf thymus (Sigma). Reconstitution conditions were as in Germond *et al.* (1976), and histone to DNA ratios were respectively (a) 1 : 1; (b) 2 : 1; (c) 2 : 1. This indicates that the presence of high amounts of the unusual base hydroxymethyluracil in dinoflagellate DNA does not impede accurate DNA-histone interactions.

bacterial basic protein (Rouviere-Yaniv and Gros, 1975; Haselkorn and Rouviere-Yaniv, 1976) has not yet been investigated. Nevertheless, we have shown, by means of *in vitro* reconstitution experiments, that purified dinoflagellate DNA was able to form nucleosomes in the presence of foreign histones (Figure 2). In addition to the evidence provided by several workers using both prokaryotic and eukaryotic DNA (Wilhelm *et al.*, 1979; Noll *et al.*, 1980) that histones, and not DNA as such, are responsible for nucleosome formation, it is now clear that high amounts of an abnormal base (hydroxymethyluracil, see below) in the DNA are not an impediment to *in vitro* construction of nucleosomes by heterologous histones.

4. Distinctive Features of Dinoflagellate Nuclei

Some of the nuclear aspects of Dinoflagellates are quite distinctive of this group. The nuclear membrane is persistent throughout the whole mitotic cycle. Chromosomes are permanently condensed and no longitudinal differentiation (e.g. Q-, G- or C-bands which occurs in eukaryotic metaphasic chromosomes after specific 'banding technique') was observed (Haapala and Soyer, 1974) in *Prorocentrum micans*. These chromosomes are characterized by a peculiar organization of the chromosomal fibers (Oakley and Dodge, 1979: Livolant and Bouligand, 1980; Herzog and Soyer, 1983), which are tightly coiled into a double helical bundle. We recently provided evidence that *P. micans* chromosomal DNA is compacted in a hierarchy of six organizational levels, based on the mode of helical coiling (Herzog and Soyer, 1983). This organization (Figure 3) allows the large amount of DNA (which in Dinoflagellates is 5 to 10 times higher than in eukaryote nuclei) to be compacted into chromosomes, in the absence of histones.

Another distinctive feature of the nuclei is the division mechanism (dinomitosis), which is still under investigation, but is known to be basically different from the eukaryote mitosis. It is mainly characterized by the absence of (a) a diffuse chromatin interphase (except for some genera like *Noctiluca, Blastodinium*) (Soyer, 1971, 1972), (b) metaphase chromosomes and metaphasic plate, (c) a *kinetochore*, with the exception of *Syndinium* and perhaps *Amphidinium* (Oakley and Dodge, 1974), *Oodinium* (Cachon and Cachon, 1979) and *Glenodinium* (Dodge, 1971), (d) centrioles, except for

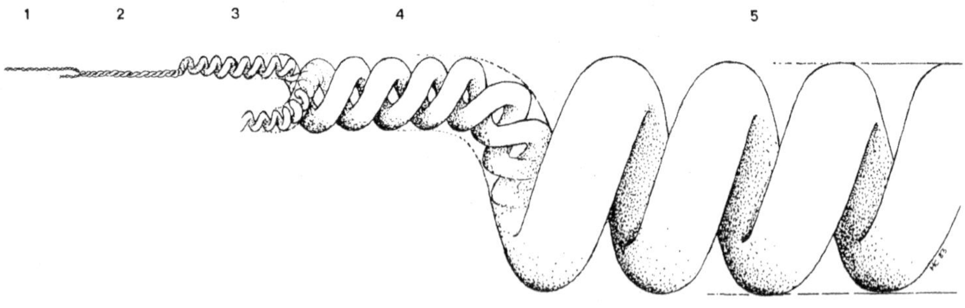

Fig. 3. Tentative representation of the helical compaction of dinoflagellate chromosomal DNA in a hierarchy of 6 organization levels. DNA molecules (level 1) form a double helix (level 2) 10 nm thick which is twisted into a single helix 18 nm in diameter (level 3). The fourth level is represented by a double-helically twisted state (25–31 nm thick) of two 18 nm filaments, which in turn are supertwisted into a left-handed single helix (level 5) with a diameter of 43–56 nm. These chromosomal fibers are finally united to form the double helical bundle of the chromosome (not shown here) with a diameter of about 1250 nm.

Syndinium (Ris and Kubai, 1974). Other features are the longitudinal division of the chromosomes which appear Y- and V-shaped, and the presence, as described in *P. micans* (Soyer, 1981 and Figure 1b), of intranuclear microcables, which might be involved in the chromosomal movement. These microcables could act in a specific mechanism of dinomitosis probably restricted to the most primitive genera.

At the molecular level, Dinoflagellates also display a peculiar trait. The abnormal behaviour of their DNA in terms of density and thermal denaturation, was shown (Rae, 1976; Herzog *et al.*, 1982) to be due to the presence of an unusual base, the 5-hydroxymethyluracil, which replaces 12–68% of the thymines in the investigated species. Figure 4 demonstrates the presence of this abnormal base in the highly complex heterotrophic species *Noctiluca miliaris*, as shown by *in vitro* labelling using *E. coli* DNA polymerase I. This base is specific to Dinoflagellates and has otherwise been found only in some bacteriophages infecting *Bacillus subtilis* (Kallen *et al.*, 1962).

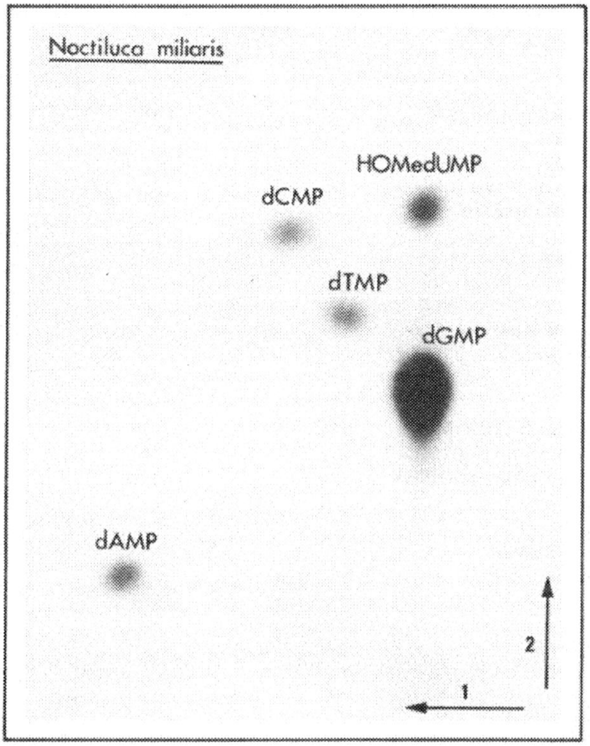

Fig. 4. The presence of the unusual nucleotide containing the base 5-hydroxymethyluracil, in *Noctiluca miliaris* DNA. Purified DNA was labelled *in vitro* by a method derived from the Nick-translation reaction, using *E. coli* DNA polymerase I, and radioactive ($\alpha\,^{32}P$) – d ATP. 3′ nucleotides were obtained by digestion of the labelled DNA by micrococcal endonuclease and phosphodiesterase II, and separated by two-dimensional chromatography. This unusual base has been found in all the so far investigated dinoflagellate species.

5. Discussion and Conclusion

The various features have been listed in Table I for comparison of dinoflagellate nuclei with both prokaryotic and eukaryotic conditions. Characters or character states were grouped using functional or structural criteria: nuclei, chromosomes, chromatin

TABLE I

Synoptic presentation of nuclear features listed for Prokaryotes, Dinoflagellates and higher Eukaryotes. Features that appear 'unique' for one group are encircled; many can be grouped according to their interrelationship. For some of these (e.g. permanence of nuclear membrane) it is impossible to decide whether they are common ancestral or uniquely derived characters (cf. Figure 5).

Character-states	Prokaryotes	Dinoflagellates	Eukaryotes
Nuclei			
Well defined nucleus	−	+	+
Nuclear membrane	−	+	+
– permanence of the membrane		+	−(a)
Chromosomes			
Permanently condensed chromosomes	−	+(b)	−
– C-, G- or Q-banding	−	−	⊞
– arch-shaped aspect	+	+	−
– helical compaction (6 levels)	−	⊕	−
– chromosome stabilizing RNA	+	+	−
Chromatin organization			
Nucleosomes	−	−	⊞
– histones	−	−	⊞
– HU-proteins (DNA-binding)	◇+	−	−
– basic non histone proteins (12–13 kd)	−	⊕	−
DNA:			
– thymine replacement by HOMeU	−(c)	⊕	−
– high amount of repetitive DNA	−	+	+
– tandemly repeated ribosomal DNA	−	+	+
Nuclear division			
– S-phase DNA synthesis	−	+	+
– Mitosis	−	−	⊞
– Dinomitosis	−	⊕	−
– Interphase		−	⊞
– Metaphase		−	⊞
– Kinetochores		−(d)	⊞
– Centrioles		−(e)	⊞
– Mitotic spindle		−(f)	⊞
– Membrane attachment of the dividing chromosomes	+	+	−
– Intranuclear microcables		⊕	−
– Nucleoli	−	+	+

(a) Except for some protozoa. (b) Except for *Noctiluca* and some parasitic species. (c) Except for bacteriophages SP 8. (d) Kinetochore-like structures in *Amphidinium* and some parasitic species. (e) Except for the parasitic *Syndinium*. (f) Extra-nuclear spindle in some species?

organization, DNA and nuclear division. This representation clearly shows the close relationship of Dinoflagellates with Eukaryotes. Nevertheless, their peculiar features described above, suggest that Dinoflagellates have diverged early from the eukaryotic lineage, as proposed by Loeblich (1976), before the eukaryotic chromatin was organized in nucleosomal structures, but after the appearance of repeated DNA sequences and, presumably, of the permanent nuclear membrane. This latter assumption is supported by the example of some lower Eukaryotes, as most of the Sporozoa, Foraminifera, Zoomastigina and Euglenophyta (for a review, see Margulis, 1981), which show a permanently intact nuclear membrane, as do Dinoflagellates, but otherwise display regular eukaryotic features.

A correlation may be seen in Eukaryotes between presence of histones and lack of arch-shape of the chromosomes. The appearance of the nucleosome structure in the higher Eukaryotes may have implied the disappearance of the structure resulting in the arch-shaped aspect of thin-sectioned chromosomes. The different structural and biochemical characters may be used to establish the cladogram presented in Figure 5, as an 'argumentation plan' *sensu* Hennig. Dinoflagellates and Eukaryotes have a common ancestor recognizable by the nuclear structure. The peculiar characters of Dinoflagellates strongly suggest that this group is monophyletic and has evolved independent of the true Eukaryotes; it thus may be considered as their sister group. This concept is in agreement with Loeblich's (1976) proposal and does not appear in any way opposed to the mesokaryotic hypothesis introduced by Dodge (1965). If future

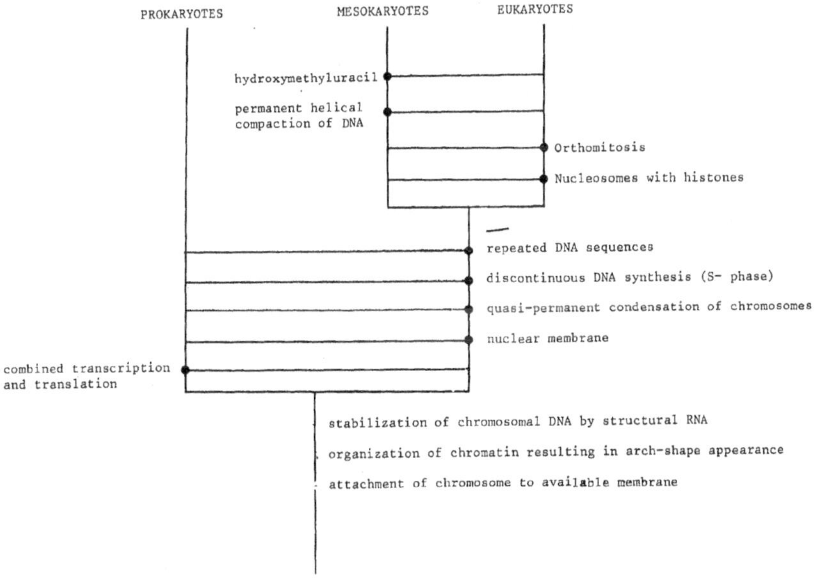

Fig. 5. A tentative cladogram based only on nuclear features supposed to present an apomorphic state (uniquely derived, or group-distinctive state) for the level considered (the corresponding plesiomorphic state on the 'sister' branch is not listed here, e.g. absence of nucleosomes).

work supports this conclusion then the current taxonomic practice of grouping Dinoflagellates with other unicellular Eukaryotes creates an unacceptable paraphyletic group.

In terms of phylogenetic systematics, the former Eukaryotes *sensu lato* would thus be subdivided into two monophyletic (sister) groups, the Mesokaryotes (or Dinokaryotes) and the Eukaryotes *sensu stricto*. However, as classification generally (and justly) lags behind the establishment of highly probable phylogenetic relationships, this nomenclature is not intended for immediate use in systematics. Rather it is proposed as a guideline for discussions focussing on the problem of dinoflagellate evolution.

References

Allen, J. R., Roberts, T. M., Loeblich, A. R., III, and Klotz, L. C.: 1975, *Cell* **6**, 161.
Bodansky, S., Mintz, C. B., and Holmes, D. S.: 1979, *Biochem. Biophys. Res. Commun.* **88**, 1329.
Britten, R. J. and Kohne, D. F.: 1968, *Science* **161**, 529.
Cachon, J. and Cachon, M.: 1979, *Arch. Protistenk.* **122**, 267.
Cavalier-Smith, T.: 1981, in Carlile, Collins and Moseley (eds), *Molecular and Cellular Aspects of the Microbial Evolution*, Cambridge Univ. Press, pp. 33–83.
Chatton, E.: 1920, *Arch. Zool. Exp. Gén.* **59**, I.
Chatton, E.: 1937, *Titres et Travaux*.
Dodge, J. D.: 1964, *Arch. Mikrobiol.* **48**, 66.
Dodge, J. D.: 1965, *Excerpta Med. Int. Congr. Ser.* **91**, 339.
Dodge, J. D.: 1971, *Protoplasma* **73**, 145.
Escaig, J., Geraud, G., and Nicolas, G.: 1977, *C.R. Acad. Sci. Paris* **284**, 2289.
Franker, C. K.: 1971, *J. Phycol.* **7**, 165.
Franker, C. K., Sakhrani, L. M., Pritchard, C. D., and Lamden, C. A.: 1974, *J. Phycol.* **10**, 91.
Galleron, C. and Durrand, A. M.: 1979, *Protoplasma* **100**, 155.
Germond, J. E., Bellard, M., Oudet, P., and Chambon, P.: 1976, *Nucl. Acids Res.* **3**, 3173.
Giesbrecht, P.: 1965, *Zbl. f. Bakt I Abt. Orig.* **196**, 516.
Gourret, J. P.: 1978, *Biol. Cellulaire* **32**, 299.
Grassé, P.-P. and Dragesco, J.: 1957, *C.R. Acad. Sci. Paris* **245**, 2447.
Grell, K. G. and Schwalbach, G.: 1965, *Chromosoma* **17**, 230.
Grell, K. G., and Wohlfarth-Bottermann, K. E.: 1957, *Z. Zellforsch. Mikroskop. Anat.* **47**, 7.
Gressel, J., Berman, T., and Cohen, N.: 1975, *J. Mol. Evol.* **5**, 307.
Haapala, O. K. and Soyer, M. O.: 1974, *Hereditas* **78**, 141.
Hamkalo, B. A. and Rattner, J. B.: 1977, *Chromosoma (Berlin)* **60**, 39.
Haselkorn, R. and Rouviere-Yaniv, J.: 1976, *Proc. Natl. Acad. Sci. U.S.A.* **73**, 1917.
Hennig, W. and Schlee, D.: 1978, *Stuttgarter Beitr. Naturk. Ser. A.* **319**, 1.
Herzog, M. and Soyer, M. O.: 1981, *Eur. J. Cell Biol.* **23**, 295.
Herzog, M. and Soyer, M. O.: 1983, *Eur. J. Cell Biol.* **30**, 33.
Herzog, M., de Marcillac, G. D., and Soyer, M. O.: 1982, *Eur. J. Cell Biol.* **27**, 151.
Hinnebusch, A. G., Klotz, L. C., Blanken, R. L., and Loeblich, A. R., III: 1981, *J. Mol. Evol.* **17**, 334.
Hinnebusch, A. G., Klotz, L. C., Immergut, E., and Loeblich, A. R., III: 1980, *Biochemistry* **19**, 1744.
Kallen, R. G., Simon, M., and Marmur, J.: 1962, *J. Mol. Biol.* **5**, 248.
Kornberg, R. D.: 1977, *Annu. Rev. Biochem.* **46**, 931.
Livolant, F. and Bouligand, Y.: 1980, *Chromosoma (Berlin)* **80**, 97.
Loeblich, A. R., III: 1976, *J. Protozool.* **23**, 13.
Margulis, L.: 1981, *Symbiosis in Cell Evolution*, W. H. Freeman and Co, San Francisco, pp. 1–419.
Noll, M., Zimmer, S., Engel, A., and Dubochet, J.: 1980, *Nucl. Acid Res.* **8**, 21.
Oakley, B. and Dodge, J. D.: 1974, *J. Cell. Biol.* **63**, 322.
Oakley, B. and Dodge, J. D.: 1979, *Chromosoma* **70**, 277.
Pettijohn, D. E., Miles, E., and Hecht, R.: 1973, *Cold Spr. Harb. Symp. Quant. Biol.* **38**, 31.
Rae, P. M. M.: 1970, *J. Cell Biol.* **46**, 107.

Rae, P. M. M.: 1976, *Science* **194**, 1062.
Reddy, R., Spector, D., Henning, D., Liu, M. H., and Busch, H.: 1983, *J. Biol. Chem.* **258**, 13965.
Ris, H.: 1962, in R. J. C. Harris (ed.), *The Interpretation of Ultrastructure*, Academic Press, New York, pp. 69–83.
Ris, H. and Kubai, D. F.: 1974, *J. Cell Biol.* **60**, 702.
Rizzo, P. J.: 1981, *BioSystems* **14**, 433.
Rizzo, P. J. and Burghart, R. C.: 1980, *Chromosoma (Berlin)* **76**, 91.
Rizzo, P. J. and Burghart, R. C.: 1982, *BioSystems* **15**, 27.
Rizzo, P. J. and Nooden, L. D.: 1974, *Biochim. Biophys. Acta* **349**, 415.
Rouviere-Yaniv, J. and Gros, F.: 1975, *Proc. Natl. Acad. Sci. U.S.A.* **72**, 3428.
Sarjeant, W. A.: 1974, in *Fossil and Living Dinoflagellates*, Academic Press, New York, pp. 13–42.
Soyer, M. O.: 1971: *Chromosoma (Berlin)* **33**, 70.
Soyer, M. O.: 1972, *Chromosoma (Berlin)* **39**, 419.
Soyer, M. O.: 1977, *Biol. Cell.* **30**, 297.
Soyer, M. O.: 1981, *BioSystems* **14**, 299.
Soyer, M. O. and Haapala, O. K.: 1974, *J. Microscopie* **19**, 137.
Steele, R. E.: 1980, Ph. D. Thesis, Yale University.
Taylor, F. J. R.: 1978, *BioSystems* **10**, 67.
Taylor, F. J. R.: 1980, *BioSystems* **13**, 65.
Wilhelm, M. L., Langenbuch, J., Wilhelm, F. X., and Gigot, C.: 1979, *FEBS Lett.* **103**, 126.

AN ESSAY ON THE EVOLUTION OF CILIOPHORAN ORAL CYTOARCHITECTURE BASED ON DESCENT FROM WITHIN A KARYORELICTEAN ANCESTRY

EUGENE B. SMALL

Department of Zoology, University of Maryland, College Park, MD 20742, U.S.A.

(Received 19 October, 1983)

Abstract. A description of how two basically dissimilar oral cytoarchitectures of ciliates, one prostomial, the other ventrostomial, might have evolved is presented. The evolutionary sequence for these types of oral organization may well have been derived from within a karyorelictean ciliate ancestry. A truly gymnostomatous condition, similar to the morphology of oral areas in extant *Tracheloraphis* species, is postulated to have preceded the evolution of oral types in which kinetids have come to support the ingestatory areas. The precursor to the *Tracheloraphis* stage is argued to have been similar to the flan-like condition of the extant ciliate genus *Kentrophoros*. Thus, it is proposed that ciliate oral architecture had its origin from within the group of karyorelictean ciliates rather than from some flagellate ancestor.

1. Introduction

In recent years, there has been a growing awareness of the existence of psammobiotic ciliates that show a diverse array of morphological types (Dragesco, 1960, 1963) most, if not all, of which also possess distinctive nondividing diploid macronuclei. These macronuclei differentiate at each division from micronuclear division products (Raikov, 1969, 1975, 1982). This has led Corliss (1974, 1979; Corliss and Hartwig, 1977) to establish a new taxon, the order Karyorelictida. Electron microscopial studies (Gerrasimova and Seravin, 1976; Nouzarède, 1977; Raikov, 1975) have revealed that the ultrastructure of the somatic kinetid is similar to that of the heterotrichine heterotrichs. Following Lynn's (1981) recent review of somatic kinetid ultrastructure, Small and Lynn (1981) have proposed a new macrosystem for the phylum Ciliophora in which karyorelictean ciliates are given class status and have been recognized as central to the evolution of the phylum (see Figure 1 in Small and Lynn, 1981).

The karyorelicteans are indeed not only karyorelicts but also, in my thinking, corticorelicts as well. Unlike other ciliates, their somatic kinetids consist of dikinetids that lack parasomal sacs. The cell membrane surface has neither the well developed alveoli typical of so many ciliates nor the perilemme of some spirotrichous groups. Contractile vacuoles (expulsion vesicles) and a well-defined cytoproct are also wanting.

Yet these same ciliates (with their unique form of nuclear duality) still possess, like other ciliates, oral attributes that run the gamut of diverse expression – all within the same class! And a whole range of feeding types may be found, filling diverse ecological niches (as pointed out by Fenchel, 1969). Oral morphological types corresponding to diverse ecological types are represented within the class by:

Origins of Life **13** (1984) 217-228. 0302-1688/84/0133-0217$01.80.

(1) forms that are only prostomatous, both raptorial and nonraptorial, within the genus *Trachelocera*;

(2) forms that have well developed ventral cytostomes bordered by heterotrich-like polykinetids (e.g., the proheterotrich genera *Avela* and *Geleia*: see Nouzarède, 1977);

(3) forms that have well developed ventral cytostomes bordered by multiple left- and right-sided oral dikinetid configurations (e.g., *Loxodes* and *Remanella*: see Dragesco, 1960; de Puytorac and Njiné, 1971); and

(4) forms, like *Kentrophoros*, that are completely astomatous, in which ingestion of attached epicommensal bacteria or diatoms takes place by simple membrane infolding (see Fenchel, 1968) in a fashion that is reminiscent of amoeboid pseudopodial engulfment.

If one excludes, for the moment, the highly evolved ciliates that have very specialized modes of ingestion – like the apostomatids, astomates, and suctoria – major oral strategies (morphologies) are obvious. One group, represented in Small and Lynn (1981) in part by the subphylum Rhabdophora, is essentially prostomial. None is truly gymnostomatous. Each has a specialized oral ciliature of oral dikinetids many of which, like the litostomean Haptoria, possess transverse microtubular ribbons orally directed on which specialized toxicysts insert. Another group, represented in Small and Lynn (1981) by the subphylum Cyrtophora, possesses ventral cytostomy, and, except for certain groups (e.g., the astomates, suctoria, etc.), has specialized oral kinetid morphologies. Fauré-Fremiet (1950) long ago recognized these two different strategies when he originally erected for the now archaic systematic group, the gymnostomes, the two major subdivisions Rhabdophorina and Cyrtophorina. Ever since reading this heuristic work, my question has been to understand the possible origin(s) of these two basically different stocks of ciliates.

Orias (1976) has proposed that the karyorelictean astome *Kentrophoros* might be looked on as primitive among all ciliates since the cell surface is organized into a flan of kinetid units on one side with ingestion taking place from the other nonciliated surface. In his view, the rounding up of the ciliated surface from the outside, thus enclosing the ingestive surfaces on the inside, would create a tube-like ciliate with digestive vacuoles located internally and somatic flan locomotory ciliature located externally. It is true that in species of the genus *Kentrophoros* there is no evidence of oral ciliature, and the whole cell surface is bound by only one cell membrane, capable of infolding to form internal cytoplasmic food vacuoles without necessarily rounding up (Raikov, 1973a, 1974). Unfortunately, Orias did not take into account two other genera of karyorelictean ciliates that are equally enigmatic: *Trachelonema* and *Tracheloraphis*.

Species of both of these two ciliate genera seem at first glance to be prostomatous since the anterior ends appear enlarged and the somatic kineties terminate in such a fashion that an anterior area, devoid of ciliature, is *almost* completely enclosed. This would suggest that these phagotrophic ciliates are prostomatous. Examination by light microscopy of diverse genera shows that, indeed, some species within the genus *Tracheloraphis* are capable of ingesting very large food substances, such as copepods (Hartwig, 1973) or large pennate diatoms (Raikov, 1962; Raikov and Kovaleva, 1968);

and a multitude of others (35 described species in the genus *Tracheloraphis* alone) are able to feed on diverse foodstuffs, including bacteria, microflagellates, diatoms, and small ciliates (Fenchel, 1968, 1969).

It is of interest to note that, in all of the published literature on such ciliates, no one has actually observed the act of ingestion itself. Dragesco, Fenchel, Raikov, and Borror (personal communications) all suggest that the ingesta of these forms may include sand grains, among which the ciliates live, which would thus make microscopical observations of the process difficult.

Frequently overlooked as an additional possible feeding surface in these organisms is the 'dorsal' glabrous zone, an area often continuous with the anterior nonciliated surface and extending to the posterior end of the cell. This zone is barren of structure, save the external limiting plasma membrane.

2. How Ciliate Oral Evolution May Have Proceeded

It has occurred to me that species of the genus *Tracheloraphis* might ingest phagotrophically from the anterior end, from the glabrous stripe, or from both. *Trachelonema* species may feed in the same fashion, since they have essentially the same cortical morphological features but are more ribbon-like and possess fewer kineties and an even broader glabrous area. Evidence in support of the glabrous zone as a feeding surface in species of the genus *Tracheloraphis* has recently been found by Lenk *et al.* (1984) working with cultures of species from North American marine and estuarine sands. When fed cooked egg yolk, these ciliates have been observed ingesting the yolk granules directly through their glabrous surfaces.

A second idea is also suggested. Ciliate prostomatous oral specializations and ventral oral specializations may well have evolved from a single *Tracheloraphis*-like ancestral stock. In the first instance, the anteriormost kinetid paratene would have become specialized for raptorial feeding from a dome-like configuration. Some extant *Tracheloraphis* species possess that kind of oral architecture, with a demonstrable 'coronne ciliaire' of oral dikinetids (e.g., see the description of *T. teissleri* in Dragesco, 1960, 1969). Observations on living species feeding prostomially have been made by but two observers, to my knowledge. Fenchel (1968) describes the process very briefly; and Spoon (personal communication) states that he has observed it, too, when some of these organisms have been feeding on small scuticociliates. Spoon's samples were taken directly from the living coral reef exhibit tanks at the Smithsonian Institution's National Museum of Natural History in Washington, D.C.

Ciliated ventral oral structures could have evolved from the glabrous stripe region of a *Tracheloraphis*-like ancestral stock, in the second instance, Protargol-stained preparations of local *Tracheloraphis* species confirm Dragesco's earlier observations that kinetosomes barren of both cilia and the microtubular postciliodesmata are to be found on both sides of the zone. In an evolutionary sense, these kinetosomes could well have given rise to the oral di- and polykinetid configurations that are found in those ciliates bearing nongymnostomatous ventral cytostomes and oral ciliature and infraciliature. This kind of regional oral differentiation would have thus produced the recognizable ventral cytostomy so characteristic of geleid and loxodid karyorelicteans. The striking ultrastructural similarity noted by Nouzarède (1977) between the ventral polymerized, densely packed serial paracytostomal monokinetids of *Avela* and *Gelia* and the serial, adoral polykinetids of heterotrichine heterotrichs further supports this contention. The left and right specialized oral dikinetids of *Loxodes* and *Remanella*, two other genera of karyorelicteans with ventral cytostomes, would lend additional support for his contention. A corollary of Lynn's (1976) structural conservatism hypothesis could be that, since dikinetids are less polymerized, less specialized than polykinetids, free-living ciliates possessing only oral dikinetids are primitive. Likewise, ciliates possessing no oral kinetid specialized structural organization would be even more primitive and thereby represent the most likely primitive stock, provided, of course, that the nonoral cortical structures were homologous and the nuclear features were homologues, as indeed they all are in karyorelicteans.

Again, *Tracheloraphis* appears to be primitive since species in the genus that have been thus far examined ultrastructurally by Raikov have no oral kinetid specialization (Raikov, personal communication with respect especially to *T. Dogieli*). In my opinion, this appears to be a truly gymnostomatous (naked mouth) condition: the total absence of orally specialized kinetids and ciliation.

The predecessor to the *Tracheloraphis-Trachelonema* prostomial and ventrostomial ciliate-like types may have been *Kentrophoros*-like. As Orias (1976) suggested, this would be a simple flat flan, ciliated on one surface only, with ingestion taking place on the nonciliated surface by membrane infolding. Contrary to Orias' point of view, however, I suggest that the next step in the sequence would not be the coiling or closing up (as *is* exhibited by *Kentrophoros fistulosum*: see Raikov, 1973a) of the nonciliated surface so that it becomes internal. In my view, *K. fistulosum* is a specialized form, since many other *Kentrophoros* species are known that also feed by presumed pseudopodial engulfment but do not coil up. Fenchel (1968) alludes to one species, found in Danish sands, that contained diatom tests. The transitional state, from a flat ribbon-like form to one that is cylindrical in cross-section, is rather to be found in a *Tracheloraphis*-like form in which the glabrous area might be considered homologous to the nonciliated surface of *Kentrophoros* and/or *Kentrophoros*-like forms. The tubular shape of *Tracheloraphis*-like species could be accounted for by increase in cell volume concomitant with increased numbers of somatic kinetid units that are aligned into kineties in such a way that the kinetid organization gradually expands to enclose an ever-increasing amount of nonciliated, glabrous cortex. Again, within the known species of the genus *Tracheloraphis*, evidence is found to support this idea:

(1) Some species with relatively low kinety numbers possess glabrous stripes that are broad, whereas other species with relatively high kinety numbers possess glabrous stripes that are relatively narrow.

(2) Branching kineties are known for at least one species (*T. striatus* Raikov, 1962). On transverse division, these incomplete kineties could become complete, thereby giving rise to progeny with an increased number of kineties.

(3) Some *Tracheloraphis* species have incomplete kineties to the left of the glabrous stripe (if the glabrous stripe, a feeding surface, can be considered ventral, rather than dorsal). Here again the numbers of kineties could increase with each transverse division that first bisected preexistent partial kineties, which were then left to replicate additional kinetids after division. Since karyorelicteans are known to recover remarkably from merotomy caused by the grinding action of sand grains in their habitats, it is certainly likely that kinetidal replications do occur with or without nuclear divisions; well recognized cortical patterns are always reproduced or reconstituted following such damage.

The next stage in this phylogenetic sequence would be to 'zipper up', or have the kineties come together in the region just posterior to or as a part of the slightly bulbous anterior end. This morphological picture *is* seen, though only in the genus *Tracheloraphis*. Complete absence of a glabrous stripe is found solely in the genus *Trachelocerca*, and all of the species described for this genus are exclusively prostomial feeders. If one turns to examine other *Tracheloraphis* species, however, the intermediate condition can be noted. In some of these forms (e.g., *T. prenanti* Dragesco, 1960, and *T. remanei* Dragesco, 1960), the anterior 'neck region' does not contain a glabrous zone but is closed by the encircling kineties.

3. Postulated Sequence of Stages

Stage 1(a). The advent of multiple kinetids in more than one kinety (see Figure 1). A single-kinety protist is not to be found in the ciliates, and one could assume that a multiple-kinetid single kinety must have originated in a flagellate ancestor. Certainly a flagellate with a highly organized cortex like that found in *Colponema loxodes* (Mignot and Brugerolle, 1976), but with nuclear endomitosis, might provide the ancestral type. Or, a simple unarmored gymnodinioid flagellate with an endomitotic nuclear apparatus is equally appealing. Replication of the nonapical kinetid in the absence of karyokinesis and cytokinesis could then give rise to a form similar to species of

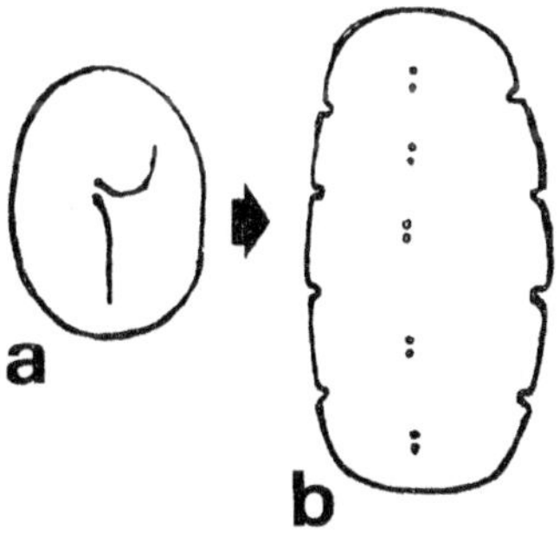

Fig. 1. The possible evolution of a single kinety with multiple paired-kinetosome kinetids from a single kinetid-bearing organism by means of kinetid replication in the absence of karyokinesis and cytokinesis.

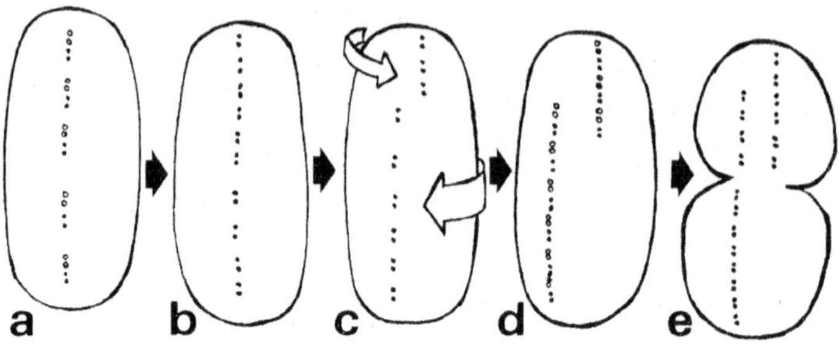

Fig. 2. A sequence of minimal steps by which an organism with but one kinety could become multikinetal. a. Replication. b. Torsion and fragmentation of the kinety. c. Further replication within each kinetal fragment. d. Perkinetal cytokinesis. e. Resultant filial cell separation in which the proter bears two kineties.

Polykrikos extant today, again with eukaryotic nuclei and endomitoses. It is interesting to note the recently published observations by Raikhel *et al.* (1981) that *Stentor*, a heterotrich in the subphylum Postciliodesmatophora of Small and Lynn (1981), has its nuclear, mitotic microtubules external to the nuclear envelope during division, thereby resembling the condition in the dividing nuclei of dinoflagellates more than it resembles the endomitosis of most other ciliates. Another important reason for selecting this hypothetical original flagellate ancestry would be that the cytokinetic fission plane would divide the cell transversally, through the replicated kinetid(s), as indeed the process occurs in extant dinoflagellates today.

State 1(b). Replication of one kinety to form multiple kineties (see Figure 2). It is known, but little recognized, in cloned ciliates studied in mass culture, that frequently unequal torsion of the anterior end of a cell (Figure 2b) may produce a spatial disjunction between the two previously connected ends of a single kinety. These kinetal fragments with their kinetids may continue to replicate kinetosomes (Figure 2c), so that by the next cell division (Figure 2d) the anterior division product (the proter) could then possess one kinety and a second kinetal fragment. If the fragment continued to replicate kinetids in the absence of all division, the ultimate outcome of this sequence of

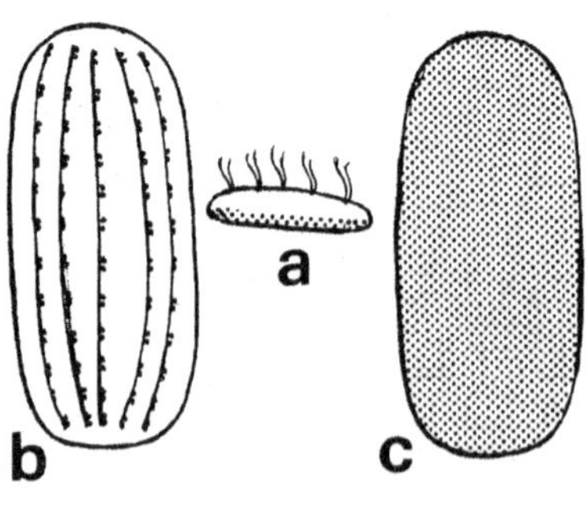

Fig. 3. The flat flan stage. a. Hypothetical cross-section in which the ingestatory surface (stippled) is separate from the kinety-bearing surface. b. Kinety-bearing surface. c. Ingestatory surface (stippled).

events would be the uneven production of multiple kineties in at least one fission product (Figure 2e). Ciliates, like the flagellate *Polykrikos*, replicate their kinetids within the kinety, and the fission plane is perkinetal in this type of transverse or oblique division.

Stage 2. The flan condition of the Orias (1976) hypothesis (see Figure 3). Continued fragmentation of a few kineties into even more numerous kinetal fragments, coupled with continued replication of the kinetids within the fragments as shown in Stage 1 (d), could result in a flan-linke organism that could have been the original *Kentrophoros*-like organism. Such an organism – to be a ciliate – would have had to meet several conditions. Firstly, this organism would have had to have a ciliate kinetid like all other ciliates, and in this primitive condition the kinetosomes would be paired within the kinetid (see Lynn, 1981; Lynn and Small, 1981; and Small and Lynn, 1981, on this point). Secondly, like *Colponema loxodes*, there would be phagocytosis of particulate food from a nonkinetidal surface, as ingestion does occur in known extant *Kentrophoros* species (Figure 3c). Thirdly, the number of kineties would be necessarily few, as is the case for extant *Kentrophoros* and *Trachelonema* species studied in recent years (see Figures 2a and 2b). No examples are known for *Kentrophoros* or *Trachelonema* with more than six kineties. Fourthly, as one finds in *Kentrophoros*, directionality of cell movement in this simple organism is determined by the way it moves and not by where its 'mouth' is. Direction of movement is again linked to the anterior-posterior orientation of the kinetids and not to a positional relationship with the oral area. Lastly, the nonkinetidal surface, the oral area that extends the length of the nonciliated cell surface, could be considered homologous to the glabrous region of *Trachelonema* and the more confined glabrous stripe of *Tracheloraphis*.

Stage 3. The *Tracheloraphis*-like stage: expansion of cell volume with increased cell biomass and increased numbers of kineties (see Figures 4a and 4b). In an evolutionary sense, a *Tracheloraphis*-like ancestral organism would have been the next step in this sequence, since at this stage one might expect to see organisms with increasing numbers of kineties. Such increased kinetal numbers could have resulted in the ribbon-flan-shaped cells becoming increasingly rounded, with the flan ingestatory surface remaining from anterior to posterior. Such *Tracheloraphis* species exist, and this morphological pattern was the very reason for Dragesco's (1960) creation of the genus *Tracheloraphis*. Again, Lenk *et al.* (1984) have observed macrophagous ingestion through this glabrous zone in several species of *Tracheloraphis*.

Stage 4. Closure of the *Tracheloraphis* glabrous stripe (see Figure 4c). Continued kinetid proliferation and increasing kinety numbers in the now *Tracheloraphis*-like tubular-body organization could, by simple constriction of the anterior region, result in two separate glabrous 'oral' areas at this stage: one would thus be anterior and prostomial and the other 'ventral' and more posterior. Again, this condition is seen in some of the known species of the genus *Tracheloraphis*. The curious apical cortical cleft, always in line with the more posteriorly separated and now ventral glabrous region, lends support to this glabrous zone closure idea (see Figures 4c, 5a and b, 5f and h).

Stage 5. Further closure of either the anterior or ventral glabrous regions so that

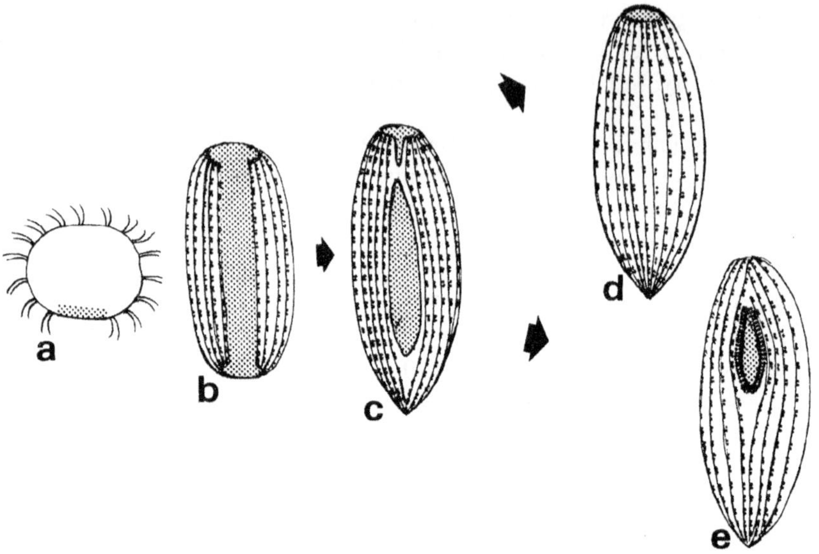

Fig. 4. The sequential steps involved in the transition from a *Tracheloraphis*-like stage to the prostomial and ventrostomial forms of more structurally complex karyorelicteans. a. A hypothetical cross-section of a *Tracheloraphis*-like form. b. A longitudinal view of a *Tracheloraphis*-like form with a continuous glabrous stripe from anterior to posterior. c. A longitudinal view of a *Tracheloraphis*-like form with a constriction of the stripe into two regions. d. A ciliate with only a prostomial ingestatory area (stippled) in which dikinetids surround the ingestatory area. e. A ciliate with only a ventrostomial ingestatory area (stippled) in which dikinetids surround the ingestatory area.

prostomial and ventrostomial ciliate stocks result (see Figures 4d and 4e). At this stage, if only the anterior glabrous region were to remain, a *prostomial* form would result. Such a condition exists in species of *Trachelocerca* (Dragesco, 1960; and see Figure 4d). Such anterior oral organization patterns of *Trachelocerca* (Figure 5l, m, and n) are perhaps not that basically different from the anterior prostomial regions of some *Tracheloraphis* forms, in which closure of the glabrous stripe within the neck region has also resulted in the apparent evolution of anterior oral specialization (see Figure 5j and k). Other *Tracheloraphis* species possess domed-like oral areas (Figures 5c, e, and h) reminiscent of the *Trachelocerca* of Figure 5m. Perhaps these are the *Tracheloraphis* and *Trachelocerca* species, noted by other authors, which feed raptorially, as do haptorid domed ciliates such as *Didinium*, *Chaenea*, and *Lacrymaria* (to mention just a few). Unfortunately, for most karyorelictean forms no ultrastructure studies have yet been carried out that would permit comparisons of their domed condition with that of the litostomial rhabdophoran species. If complete closure of the anteriormost region of this tubular *Tracheloraphis*-like primitive stock occurred, a ciliate with a *ventral* oral area would be the result (see Figure 4e). Coincident with this closure, the evolution or differentiation of specialized oral kinetids could have evolved.

Known representatives with ventral oral architecture are represented by several

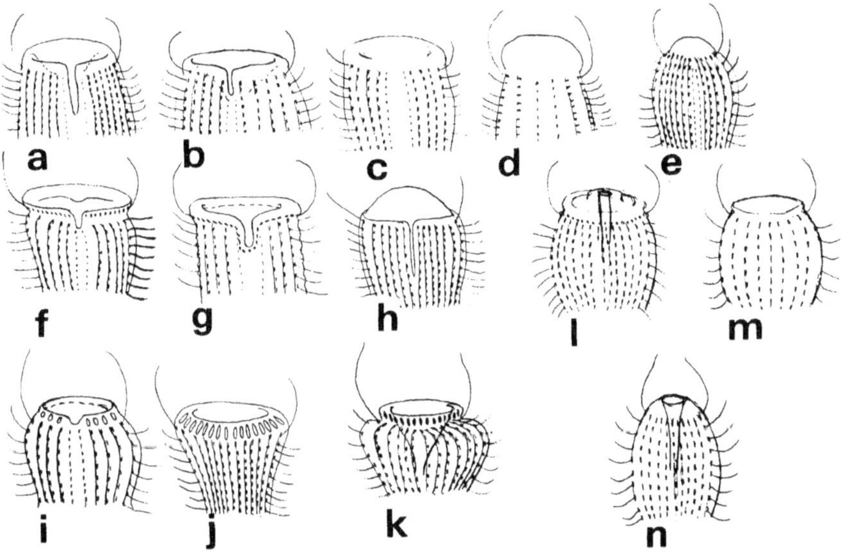

Fig. 5. The anterior ends of some *Tracheloraphis* and *Trachelocerca* species illustrating some of the diversity in presumed 'oral' morphology. a.-k., *Tracheloraphis* spp.; l.-n., *Trachelocerca* spp. a. *T. dogieli* (after Dragesco, 1963). b. *T. margaritaceus* (after Dragesco and Raikov, 1966). c. *T. prenanti* (after Dragesco, 1960). d. *T. bodiani* (after Dragesco, 1963). e. *T. stephani* (after Dragesco, 1965). f. *T. hyalinum* (after Dragesco, 1960). g. *T. teisserieri* (after Dragesco, 1963). h. *T. phoenicopterus* (after Raikov and Kovaleva, 1978). i. *T. remanei* (after Dragesco, 1960). j. *T. incaudatus* (after Dragesco, 1960). k. *T. swedmarki* (after Dragesco, 1960). l. *Trachelocerca entzi* (after Dragesco, 1960). m. *T. binucleata* and n. *T. lacrylmariae* (after Dragesco, 1960) m. The oral cleft, visible in Figures a, b, c, f, g, h, and i, always appears anterior to the glabrous stripe. Some species in both genera lack the cleft. A presumably domed oral configuration can be seen in both *Tracheloraphis* spp. d, e, and h, and in *Trachelocerca* sp. m. Likewise, a prostomial cytostome-cytopharynx is observable in *Tracheloraphis* sp. k and *Trachelocerca* spp. l and n. These arrays of prostomial morphology suggest to the author that the prostomial ciliates shared a common ancestry with these karyorelicteans.

genera of karyorelicteans, each of which can be characterized, in part at least, by its oral kinetal morphology. See Figures 6a and b of *Loxodes* and *Cryptopharynx*, respectively; both have oral dikinetids adjacent to or bordering the oral regions. Curiously, in these two genera, oral dikinetids are to be found in the area that one would think of as the cytopharynx, an internal cell surface. The structural, functional, and developmental significance of this type of cytopharyngeal kinetid remains unknown. Other karyorelictean genera with multiple-kinetosome kinetids to the right, left, anterior, or posterior in position with respect to the cell 'mouth' are illustrated in Figures 6c, d, and e. Figure 6c is a figure of a *Remanella* species with polykinetids to the left and a serial oral dikinetid to the right of and anterior to the cytostome; this is morphologically similar to the oral architecture of *Loxodes*, as illustrated in Figure 6a. Figure 6d is a figure of the oral region of a species of *Protocruzia* with six preoral serial polykinetids and a paroral, right dikinetid.

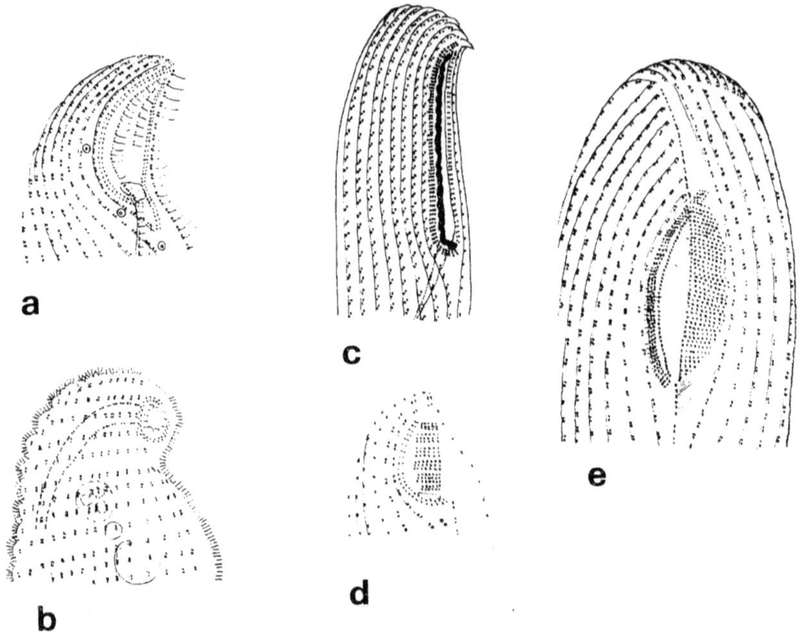

Fig. 6. Examples of karyorelictean ciliates with ventral oral configurations: a. *Loxodes vorax*, multiple and oral dikenetid both to the left and to the right, preoral and endoral kinetal fragments. b. *Cryptopharynx* sp., perioral and endoral kinetal fragments. c. *Remanella* sp. Right side, single oral polykinetid; left side, single oral dikinetid segment. d. *Protocruzia* sp., six preoral serial polykinetids, right oral dikinetid, and single postoral dikinetid kinetosomal pair. e. A geleiid sp., right side, oral monokinety and polykinety; left side, single oral polykinety. Note the absence of postciliodesmata from pre- and post-oral 'somatic' kinetids and the presence of presumed nematodesmata underlying the right oral polykinetid and the postoral 'somatic' dikinetal fragment.

Figure 6e is a figure of a species of a new geleiid genus found in the Chesapeake Bay estuary. Like species of other geleiid genera, it possesses polykinetids to the right and to the left of the ventral oral region. Preliminary observations on the feeding behavior of live organisms of this new form and related *Geleia* species (see Lenk and Small, 1983) have shown that these ciliates can and do actively ingest large cooked egg yolk particles through this oral region, in a manner quite similar to the ingestion of the same foodstuff, through the glabrous stripe in species of *Tracheloraphis* (Lenk *et al.*, 1984). The difference between the two forms is the confinement of the area in the former and the addition of oral ciliature in the latter. These observations are noteworthy in light of the ultrastructural similarity noted by Nouzarède (1977) between the left oral single polykinetid of *Geleia* and *Avela* and the multiple oral polykinetids of heterotrichine heterotrichs. The actual feeding mechanisms, however, do not appear to be the same. Clearly an additional evolutionary morphological step was taken in the evolution of heterotrich spirotricheans from geleiid proheterotrichid karyorelicteans, both considered subgroups of the subphylum Postciliodesmatophora (Small and Lynn, 1981, 1983).

4. Conclusion

An account of how two basically dissimilar oral morphological strategies might have evolved within a karyorelictean ancestry from a single karyorelictean stock of flan-like ciliates has been presented. Supporting the overall idea is our laboratory observation of macrophagous particle-ingestion by a ventral nonciliated region, the glabrous stripe, in species of *Tracheloraphis*. The present hypothesis of ciliate evolution, while utilizing the flan condition, a term and concept suggested by Orias (1976), differs from that worker's by not considering the 'rolling up' process of a few *Kentrophoros* species as crucial to the entire evolutionary sequence. Rather, the more generalized body plans of extant karyorelicteans are here considered to reflect the early evolutionary process by which both prostomial and ventrostomial cytoarchitectures could have evolved within the Karyorelictea.

Beyond the subphylum Postciliodesmatophora, two other subphyla (ranks accorded by Small and Lynn, 1981, 1983) are to be found: the Rhabdophora, with essentially prostomial cytoarchitectural morphologies; and the Cyrtophora, with essentially ventrostomial cytoarchitectural morphologies. Both of these subphyla could have evolved their oral apparatus, as well as their complex nuclear conditions, from a single karyorelictean stem group. A remaining challenge is to find a rational explanation for the transition to the more structurally conservative somatic kinetid states of the six ciliate classes (two rhabdophoran and four cyrtophoran) of these two subphyla. It is possible that continued ultrastructural analyses of both cortical and oral morphologies of the diverse karyorelictean species may supply answers to these intriguing unresolved questions as well as answers to questions that have not yet been imagined.

Ciliate 'oralness', as overwhelmingly diversified as it first appears to the student of the Ciliophora, can possibly be explained, then, as having originated from within a karyorelictean ancestral group. I have postulated that this stem group was initially gymnostomatous and that it preceded kinetal elaboration and differentiation to evolve into the two major oral lineages recognized today within the total ciliate assemblage. These two lineages, prostomial and ventrostomial, have extant examples in both karyorelictean and nonkaryorelictean taxa.

References

Corliss, J. O.: 1974, *J. Protozool.* **21**, 207–220.

Corliss, J. O.: 1979, *The Ciliated Protozoa: Characterization, Classification, and Guide to the Literature*, 2nd ed., Pergamon Press, London and New York.

Corliss, J. O. and Hartwig, E.: 1977, *Mikrofauna Meeresboden* **61**, 65–88.

Dragesco, J.: 1960, *Trav. Stat. Biol. Roscoff (N. S.)* **12**, 1–356.

Dragesco, J.: 1963, *Cah. Biol. Mar.* **4**, 91–119.

Dragesco, J.: 1965, *Cah. Biol. Mar.* **6**, 357–399.

Fauré-Fremiet, E.: 1950, *Bull. Soc. Zool. Fr.* **75**, 109–122.

Fenchel, T.: 1968, *Ophelia* **5**, 73–121.

Fenchel, T.: 1969, *Ophelia* **6**, 1–182.

Gerrasimova, Z. P. and Serravin, L. N.: 1976, *Zool. Zh.* **55**, 645–656. (in Russian with English summary)
Hartwig, E.: 1973, *Mikrokosmos* **62** (11), 329–336.
Lenk , S. and Small, E. B.: 1983, *J. Protozool.* **30**, 23A.
Lenk, S., Small, E. B., and Gunderson, J.: 1984, *Origins of Life* **13** (this issue).
Lynn, D.: 1976, *J. Protozool.* **23**, 302–314.
Lynn, D.: 1981, *Biol. Rev.* **56**, 243–292.
Lynn, D. and Small, E. B.: 1981, *BioSystems* **14**, 377–385.
Mignon, J.-P. and Brugerolle, G.: 1976, *Protistologica* **11** (year 1975), 429–444.
Nouzarède, M.: 1977, *Bull. Stat. Biol. Arcachon* (N. S.) **28** (Suppl.(year 1976), 1–315.
Orias, E.: 1976, *Trans. Amer. Micros. Soc.* **95** , 415–429.
Puytorac, P. de and Njiné, T.: 1971, *Protistologica* **6** (year 1970), 427–444.
Raikhel, N., Paulin, J. J., and Skarlato, S. A.: 1981, *J. Protozool.* **28**, 103–107.
Raikov, I. B.: 1962, *Cah. Bio. Mar.* **3**, 325–361.
Raikov, I. B.: The Macronucleus of Ciliates, in Tze-Tuan Chen (ed.), *Research in Protozoology* **3**, Pergamon Press. London and New York, pp. 1–128.
Raikov, I. B.: 1973a, *Ann. Stat. Biol. Besse-en-Chandesse* **6-7** (years 1971-72), 21–54.
Raikov, I. B.: 1973b, *Tsitologiya* **15**, 144–149. (in Russian with English summary)
Raikov, I. B.: 1974, *Cah. Biol. Mar.* **15**, 379–393.
Raikov, I. B.: 1975, *Tsitologiya* **17**, 739–747. (in Russian with English summary)
Raikov, I. B.: 1982, *The Protozoan Nucleus: Morphology and Evolution*, Springer-Verlag, Vienna and New York.
Raikov, I. B. and Kovaleva, V. G.: 1968, *Acta Protozool.* **6**, 309–333.
Small, E. B. and Lynn, D.: 1981, *BioSystems* **14**, 387–401.
Small, E. B. and Lynn, D.: 1983, Ciliophora, in J. J. Lee, S. H. Hutner, and E. C. Bovee (eds.), *Illustrated Guide to the Protozoa*, Society of Protozoologists, Allen Press, Lawrence, Kansas. (in press)

PRELIMINARY OBSERVATIONS OF FEEDING IN THE PSAMMOBIOTIC CILIATE *TRACHELORAPHIS*

SUSAN E. LENK, EUGENE B. SMALL, and JOHN GUNDERSON
University of Maryland, Department of Zoology, College Park, MD 20742, U.S.A.

(Received 19 October, 1983)

Abstract. Feeding on large particulates in *Tracheloraphis* species has been observed in living specimens to occur along the nonciliated glabrous stripe rather than at the narrow anterior end. The absence of specialized oral kinetids in this region indicates that feeding occurs by some form of membrane infolding. Organisms were cultured using cooked egg yolk as the sole source of added nutrient, enabling the observation of the feeding process in these relatively primitive ciliates.

1. Introduction

The genus *Tracheloraphis* Dragesco is a primitive genus in a group of psammobiotic ciliates (Corliss, 1975; Corliss and Hartwig, 1977) within the newly erected class Karyorelictea (Small and Lynn, 1981, 1983). These organisms are found widely distributed in both marine and estuarine habitats. A major distinguishing characteristic of the genus is the presence of a nonciliated glabrous stripe, one to several kinetal rows in width, which extends the complete length of the cell in most of the presently described species. In *Tracheloraphis dogieli* (up to now, the only *Tracheloraphis* species whose cytoarchictural ultrastructure has been characterized with transmission electron microscopy), a single row of nonciliated paired kinetosomes are found on either side of the stripe (Raikov *et al.*, 1975). In protargol-stained specimens of other species, barren (without cilia) paired kinetosomes also may be seen.

Direct observation of feeding in tracheloraphid ciliates is undescribed. The assumption of apical feeding has been based on the presence of cytologically unknown inclusion bodies grouped together at the anterior end and of long flexible cilia around the invaginated anterior end of some but not all species (Dragesco, 1960). Their somewhat *Lacrymaria*-like shape may have also influenced the notion of similar modes of feeding by members of the two genera. In fact, observations of protargol-stained whole specimens by Dragesco (1963) and by us (unpublished), of the electron micrographs of *T. dogieli* (Raikov *et al.*, 1975), and of Raikov's (see Raikov and Kovaleva, 1968) published drawings of other nonprotargol stained specimens from Eurasia have yielded neither consistent nor sufficient information to state that a true ciliate cytostomal apparatus does or does not exist at the apical end of tracheloraphids (see Small, 1983, for further clarification of this important point). This anterior region has, nevertheless, been described as the oral area (Dragesco, 1960; Raikov *et al.*, 1975). A few instances of large particulate foodstuffs in presumed food vacuoles have been noted: for example, diatoms (Raikov and Kovaleva, 1968), and a copepod, copepod egg case, and a foraminiferan (Hartwig, 1973). No explanation has been given for their

Origins of Life **13** (1984) 229-234. 0302-1688/84/0133-0229$00.90.

presence or for their mode of entry into the cell. We have observed large inclusions such as centric and pennate diatoms, cyanobacterial filamentous trichomes, and one large undetermined ciliate. These observations suggest that feeding might occur at a site away from the narrow anterior end. Feeding in these organisms might occur along the glabrous stripe by some ingestatory process based on the presence of these very large particulates enclosed in presumed food vacuoles and the presence of only a single cell membrane enclosing the glabrous stripe.

Using a modification of a cooked egg yolk medium devised by one of us (J.G.), *Tracheloraphis* species were grown for the first time in culture. The use of this culture technique enabled us to observe the feeding process.

2. Methods

Tracheloraphis species were isolated from sediments at three sites in the mid-Atlantic region: two estuarine strains from the Chesapeake Bay – Point Lookout, MD and Gloucester, VA – and the other derived from full-strength sea-water from Chincoteague, VA. Initially, the tracheloraphids used were separated from their sandy sediments by means of the Uhlig (1966) technique, as modified by Fenchel (1969), and later by Boynton and Small (1983). A 35 μm nylon monofilament mesh screen (Nitex, from Tobler, Ernst, and Traber, Inc.), separated the original sand-water mixture from the water into which the ciliates moved via the above separation technique. Otherwise, the derived organisms were concentrated by mouth pipette, picking them up from petri dishes that contained small amounts of the sand-water mixture diluted with water from each of the collection sites. Both of these techniques required the use of a binocular compound stereo-microscope (Bausch and Lomb, Inc.) whose magnification ranged from 20–90 diameters. Following extraction and isolation, the ciliates were starved for 24 hours before transfer by mouth pipette to the culture medium.

The culture medium consisted of 0.45 μm millipore-filtered water from each of the collection sites to which streptomycin-SO_4 and penicillin-g were added so that a final concentration of 100 mg/l was achieved for each (Kohlmeyer and Kohlmeyer, 1976). Nutrient was supplied by the addition of very small amounts of freshly boiled egg yolk (Balamuth and Sandza, 1944). A few particles were added to the culture dishes every 5–7 days. Subculturing was effected after a 2–4 week acclimation period.

Light microscopy was carried out using a Nomarski interference microscope (Zeiss, Inc.). Photomicrography of the feeding process was accomplished by using a 35 mm attachment camera back (Exacta) on a trinocular nosepiece. In conjunction with a substage microflash equipped with a 1/2000 second flash xenon light source (Zeiss, Inc.), 200 ASA high speed color negative film (Kodak Ektachrome) was used.

3. Results

Feeding on egg yolk particles occurs 1–3 hours after starved organisms are transferred to the medium. Feeding takes place not at the anterior end but mid-ventrally along the

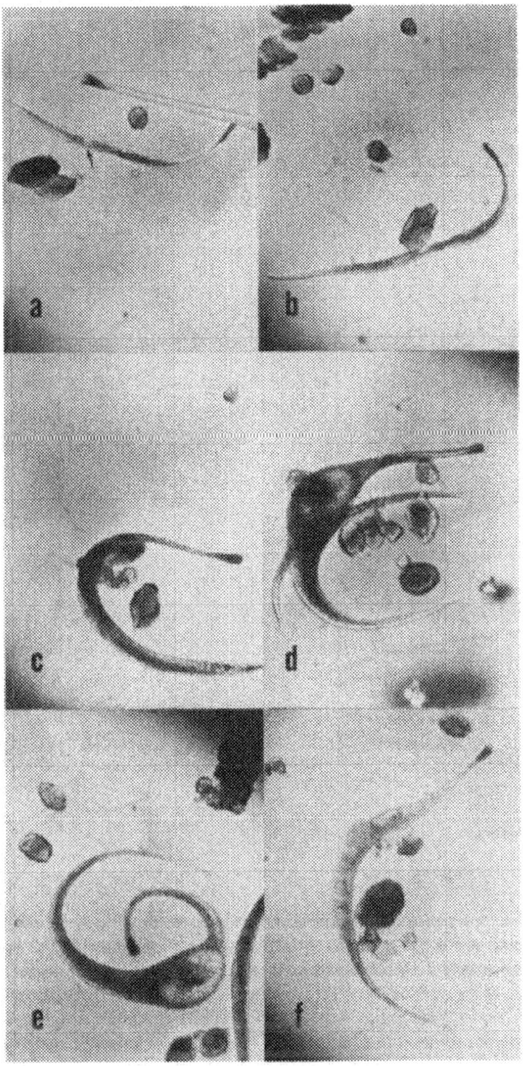

Fig. 1. Feeding in *Tracheloraphis* species. a. Two freshly isolated tracheloraphid ciliates. Arrow indicates single nuclear complex. b. Ciliate 'rubbing' and twisting its body along the surface of an egg yolk particle, with the glabrous stripe touching the particle surface. c. Onset of particle ingestion. d. Partially ingested egg yolk particle. The presumptive food vacuole has formed, and the body has become enlarged at the site of ingestion. e. Particle fully enclosed in the food vacuole. f. Ciliate with multiple food vacuoles containing egg yolk particles.

glabrous stripe (Figures 1 and 2a-f). In forms with a single nuclear complex, feeding usually occurs posterior to that complex. No observations have been made of the process in cells that have multiple nuclear configurations.

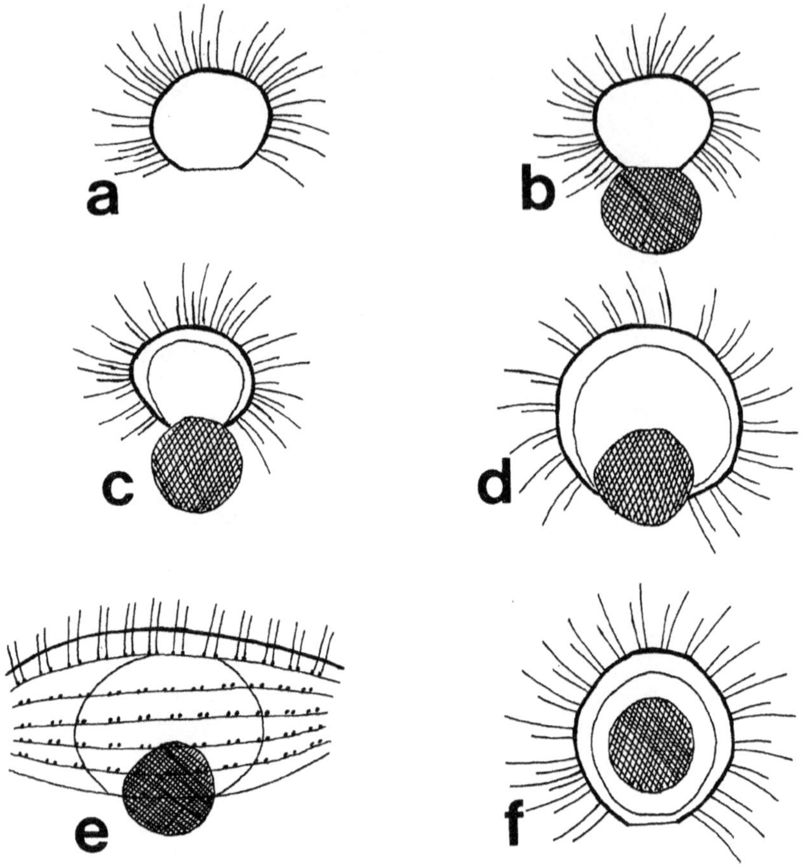

Fig. 2. Diagrammatic representation of the feeding sequence for *Tracheloraphis* species based on living specimens. a. Cross-section of a hypothetical tracheloraphid ciliate showing the nonciliated glabrous stripe (lower surface). b. Initial contact with the particle at the ingestatory site. c. Onset of particle ingestion with the organism attached to the particle. d. Partial ingestion of the particle showing the formation of the presumptive food vacuole and the enlargement of the body. e. Lateral view of the organism showing particle ingestion and presumptive food vacuole formation. f. Particle completely enclosed in a food vacuole, and reformation of the glabrous region.

Prior to feeding, the ciliates may or may not 'bump' the particle with the anterior end. This activity, if present, is followed by an action whereby the organisms appear to twist the glabrous stripe from one side to the other and, in the process, make contact with the particle surface (Figure 1b). The particle may also touch the somatic cilia of the somatic kineties on either side of the stripe. The organism so orients its body that it is atop the particle with the glabrous stripe flush against the particle surface (Figure 1c). During this time, a presumptive food vacuole is being formed and may ultimately achieve a size much larger than the food particle, prior to the actual ingestion of it

(Figure 1d). The ciliate appears to 'push' against the piece of food that is lying on the petri dish bottom until the particle is fully enclosed in a food vacuole. Then the ciliate glabrous stripe appears to be continuous again (Figure 1e).

Ingestion of large particulates results in the deformation of the body at the site of ingestion with the anterior and posterior regions not visibly affected. The ciliates do not appear to stop feeding after the ingestion of a single bit of egg yolk, as proven by the presence of large numbers of food vacuoles in some organisms (Figure 1f). The particles found in any one vacuole may vary in both size and number, with a large vacuole containing from one to six particles of varied sizes.

Observations have been made of the organisms expelling diatom frustules, after which the ciliates return to their original shape. They have not been seen to expel the egg yolk particles. It should be noted that these organisms lack any form of organized cytoproct (Raikov *et al*., 1975, and our own observations based on stained preparations). We do not yet know how elimination of particulate waste substances occurs in these psammobiotic protozoa.

4. Discussion

Feeding on large particulates in *Tracheloraphis* species occurs along the nonciliated glabrous stripe rather than at the narrow anterior end. We hypothesize that this phagotrophic process may be a form of membrane infolding, which differs from phagocytosis in rhizopod amoebae in that no pseudopodial food cup is formed (Marsot and Couillard, 1978). In the absence of transmission electron micrographic evidence, we cannot comment at this time on the exact mechanism by which the food vacuole forms from the cell membrane itself or from precursory flattened vesicles added to the site of ingestion. Unlike the more evolutionarily advanced ciliate groups, true oral dikinetidal architecture is absent (see Small, 1983) nor is there a single site of ingestion. Feeding appears to be possible anywhere along the glabrous stripe. This mode of ingestion has not been demonstrated in any of the other described karyorelictean ciliate genera (see Fenchel, 1968) that possess similar glabrous striped surfaces. Such studies should be made to more completely understand the broader significance of these preliminary observations of ours.

At present the taxonomy of these organisms is in a state of anarchy due to the absence of decently stained organisms and subsequently detailed comparative analyses of their cytoarchitecture. The cells are highly contractile and fragile (Dragesco, 1960). In addition, various nonkinetosomal cytoplasmic inclusions (e.g. various presumed extrusomes), are stained – when the protargol silver technique is used – making it impossible to accurately determine kinetal organization, believed to be crucial to systematic assignment (Small and Lynn, 1983). Many of the known species are thus far poorly studied cytologically and identifications are based on either live observations or from insufficiently stained preparations. Prior to this study, no one has successfully cultured any species of the genus.

Dragesco (1960) suggested that toxicysts (formerly called toxic trichocysts) were involved in food capture. However, in our stained and sectioned preparations there has been no supportive evidence for this conjecture. Dragesco (1962) further studied predation by *Dileptus* and other rapacious haptoreans, showing that toxicysts are involved in prey capture in those cases. However, no cytological evidence is available to support the notion that *Tracheloraphis* is rapacious. Lysis of the egg yolk particle does not occur until long after the particle has been enclosed in the food vacuole. Like *Dileptus*, food vacuole formation appears to occur during particle ingestion. Exactly how tracheloraphids capture copepods and other large prey is uncertain. Perhaps these prey are already moribund. The long anteriormost cilia may be sensory, although there is no evidence to support this, other than the occasional contact with food particles prior to ingestion.

Preliminary, yet direct, observations of macrophagous ingestion in *Tracheloraphis* species has been observed, and the uniqueness of the process, as far as we know it, has been critically appraised. This paper offers the suggestion that feeding is not prostomeal in at least the organisms we have thus far observed. Further, the glabrous stripe appears to be the site for the ingestion of food. In the absence of a distinct cytostome in the glabrous region, some form of folding-in of the glabrous zone membrane must indeed *initially* take place. Further studies of the feeding to explain more mechanistically the role of membrane precursors in the ingestatory process at the electron microscopic level are under way by one of us (S.E.L.).

References

Balamuth, W. and Sandza, J. G.: 1944, *Proc. Soc. Exper. Biol. Med.* **57**, 161–163.

Boynton, J. E. and Small, E. B.: 1983, *Sci. Teacher* (in press).

Corliss, J. O.: 1975, *Trans. Amer. Microsc. Soc.* **94**, 224–267.

Corliss, J. O. and Hartwig, E.: 1977, *Mikrofauna Meeresboden* **61**, 65–88.

Dragesco, J.: 1960, *Trav. Stat. Biol. Roscoff* (N.S.) **12**, 1–356.

Dragesco, J.: 1962, *Biol. Bull. Fr. Belg.* **96**, 123–166.

Dragesco, J.: 1963, *Cah. Biol. Mar.* **4**, 91–119.

Fenchel, T.: 1968, *Ophelia* **5**, 73–121.

Fenchel, T.: 1969, *Ophelia* **6**, 1–182.

Hartwig, E.: 1973, *Mikrokosmos.* **62**, 329–336.

Kohlmeyer, J. and Kohlmeyer, E.: 1979, *Marine Mycology: The Higher Fungi*. Academic Press, New York, 690 pp.

Marsot, P. and Couillard, P.: 1978, *Can. J. Zool.* **56**, 1497–1506.

Raikov, I. B. and Kovaleva, V. G.: 1968, *Acta Protozool.* **6**, 311–333.

Raikov, I. B., Gerassimova-Matvejeva, Z. P., and Puytorac, P. de: 1975, *Acta Protozool.* **14**, 17–42.

Small, E. B.: 1984, *Origins of Life* **13** (this volume).

Small, E. B. and Lynn, D. H.: 1981, *BioSystems* **14**, 387–401.

Small, E. B. and Lynn, D. H.: 1983, The Ciliophora, in J. J. Lee, S. H. Hutner, and E. C. Bovee (eds.), *Illustrated Guide to the Protozoa*, Society of Protozoologists, Allen Press, Lawrence, Kansas (in press).

Uhlig, G.: 1966, *Zool. Anz.* (Suppl.) **29**, 151–157.

METHODS OF SYSTEMATIC ANALYSIS: THE RELATIVE SUPERIORITY OF PHYLOGENETIC SYSTEMATICS

DIANA L. LIPSCOMB

Department of Biological Sciences, George Washington University, Washington, D.C. 20052, U.S.A.

(Received 19 October, 1983)

Abstract. The superiority of cladistic methods to both synthetic and phenetic methods is briefly advanced and reviewed. Cladistics creates testable hypotheses of phylogeny that also give a highly informative summary of available data. Thus it best fits the criteria for a method for determining the general reference classification in biology.

For protistologists in particular, cladistics is especially useful. Inundated by an abundance of ultrastructural, biochemical, and cell biological information, protistologists could be greatly helped by the informative way in which cladistics orders and summarizes the data. In addition to classifying protist taxa, hypotheses about the evolution of cell organelles and cellular could be scientifically formulated and tested by cladistics. Because cladistic classifications best summarize the data, they would also be best for making predictions about taxa and characters. They would, for the same reason, be the most stable. Widespread adoption of cladistic methods would serve to stabilize the now fluid state of protist taxonomy. It is for all of these reasons that such methods best suit the needs of the evolutionary protistologist.

1. Introduction

Classification is the ordering of organisms into groups on the basis of their relationships (Simpson, 1961, as modified by Sneath and Sokal, 1973). In the current taxonomic literature, proponents of the three main approaches to classification are engaged in a lively debate over which method will produce the best classification. These approaches are known as synthetic or traditional systematics, phenetic or numerical taxonomy, and cladistic or phylogenetic systematics.

The underlying philosophies and methods of the various schools are often difficult to grasp because of the host of papers debating methods and concepts using specialized terminology and seemingly obscure arguments. In this paper, I shall attempt to clarify the philosophies and methods of the three main taxonomic schools and discuss why I believe that the cladistic method is the best for classifying the unicellular eukaryotic organisms (the protists). Although obviously not exhaustive, my review is intended to provide protistologists with both an explanation of the differences among the synthetic, phenetic, and cladistic approaches and a background for understanding taxonomic papers that have utilized specialized applications or individual variations of the methods.

I shall not discuss all the aspects of taxonomic research. Taxonomists must choose taxa, investigate characters, and assign character states, as well as classify organisms into groups. The debate among synthetic, cladistic, and phenetic taxonomists concerns this last problem – how to best group organisms into taxa. Therefore, I shall restrict this paper to that aspect of classification and shall not discuss the varied methods of collecting characters, assigning character states, or choosing taxa.

Origins of Life **13** (1984) 235-248. 0302-1688/84/0133-0235$02.10.

2. Systematic Philosophies and Methods

2.1. Synthetic taxonomy

Advocates of the traditional synthetic taxonomic methods, as defined by Mayr (1969) and Simpson (1961), contend that hypothetical reconstructions of evolutionary history are possible and basic to classification. Major criteria for grouping organisms are common ancestry and weighted similarity. Based on expert knowledge of the animals and an understanding of evolutionary principles, the taxonomist constructs a story about the evolution of the group, concludes which characters are homologous, and groups organisms together by a differential weighting of these characters so as to reflect the evolutionary story.

This method is not repeatable since it is based on a story of evolution that involves opinions of the taxonomist with respect to what might have taken place. Belief in a different story might well result in a different classification. For example, Lynn (1979a) has argued that because different oral structures of ciliates are useful in feeding on different food types, natural selection could select for oral polymorphism (diverse morphology of the oral architecture) within a group of ciliates to take advantage of different types of food available. Natural selection would not so favor polymorphism in somatic kinetid infrastructure and it would thus be conserved. He argues further that differences in such infrastructure should correlate with evolutionary branching points. So Small and Lynn (1981) have constructed a classificiation of the ciliates placing heavy weight on somatic kinetid ultrastructural characters and relatively little weight on oral ultrastructural characters.

Another taxonomist might note that, during the morphogenesis of cell division in ciliates, one or more kinetids of the mother cell become totipotent – that is, they give rise to the different oral, somatic, and specialized kinetid infrastructures. This taxonomist might conclude that a classification should not be based on the somatic kinetid infrastructure, because it has the potential to change easily as shown by its ability to change during morphogenesis. This hypothetical taxonomist would give heavier weight to some other features that he felt were more stable – morphogenetic data, for example.

Such arguments are common in synthetic systematics, as one taxonomist tries to convince another of the probability of his evolutionary story. Ideally, the conflict among the opposing viewpoints should stimulate research that would generate the data to resolve the conflict and produce a universally accepted story. This method is widely used by protistologists although it has never been philosophically defended as the best method for classification of protists.

2.2. Phenetic taxonomy

Phenetics (numerical taxonomy) was developed in an attempt to remove conjecture and taxonomic compromise from classification and make it more repeatable and scientific. Pheneticists work from the premise that evolutionary history is unknowable and, therefore, cannot be used as the basis for classification (Sneath and Sokal, 1973).

The principle criterion for grouping and ranking taxa in phenetics is overall similarity. All characters are weighted equally (i.e., each is considered to be of equal significance). The taxonomist collects data and codes them either as present (+) or absent (−), as states (0, 1, 2, etc), or degrees (0, .5, .67, 1.2, etc). The character states for each taxa are compared to those of every other taxon. A measure of similarity (often expressed as percent of character states shared) among all the taxa is calculated. Organisms with highest overall similarity are grouped most closely together.

In protistology there have been only a few phenetic analyses (e.g., Friz (1979) on sarcodinids; Holt *et al.* (1974), and Lynn (1979b), on ciliates; and Van Valkenburg *et al.* (1975), on photosynthetic plankton). But there have not been any discussions of phenetic philosophy in the protistological literature or arguments advocating this method over others.

2.3. Cladistic taxonomy

Cladistics (phylogenetic systematics), as it was developed by Hennig (1950, 1966), was also created to make classification more repeatable and scientific. However, unlike phenetic taxonomists, advocates of cladistic classification contend that testable hypotheses of phylogeny can be generated. Ultimately, these hypotheses should serve as the basis for classification. Cladistic taxonomists group organisms together based on shared derived characters (synapomorphies) rather than on degree of overall similarity and divergence. Derived character states are distinguished from primitive character states (plesiomorphies) by a variety of methods of which out-group comparison is considered to be the most powerful (Farris, 1982; Lundberg, 1972; Stevens, 1980). In out-group comparison, character states are compared to those in a taxon outside (the out-group) the group being classified. If the state is the same as that in the out-group, then it is considered to be a primitive trait retained from the time of the common ancestor of the two groups. If the states are different, then it is considered to be a unique (derived) character state.

Inherent in the concept of out-group comparison is the parsimony criterion (Lunberg, 1972). The principle of parsimony is widely used in science and is often referred to as Occam's (or Ockham's) razor, after the 14th century philosopher and theologian William of Ockham in Surrey, England (Copeston, 1972). He wrote, "Nunquam ponenda est pluritas (sine necessitate)", and "Frustra fit per plura quod potest fieri per pauciora", which translate to "Never is multiplicity to be postulated without necessity" (Kneale and Kneale, 1962, p. 243), and "It is vain to do with more what can be done with fewer" (Russell, 1946, p. 494). In other words, the most parsimonious or simplest hypothesis is the one that requires the fewest *ad hoc* statements to explain the full array of data. And such a hypothesis should be preferred over others that compete for the same idea.

The out-group method always puts the root on a cladogram so as to achieve the most parsimonious connection with organisms outside the studied group (Farris, 1980; Lundberg, 1972). Figure 1 illustrates an example of why this is so. Figure 1a shows the distribution of the character states of five characters for three taxa (A, B, and C). Two of

Characters	Taxa		
	A	B	C
I	b	a	a
II	a	b	a
III	a	a	b
IV	b	b	a

	Outgroup
I	a
II	a
III	a
IV	a

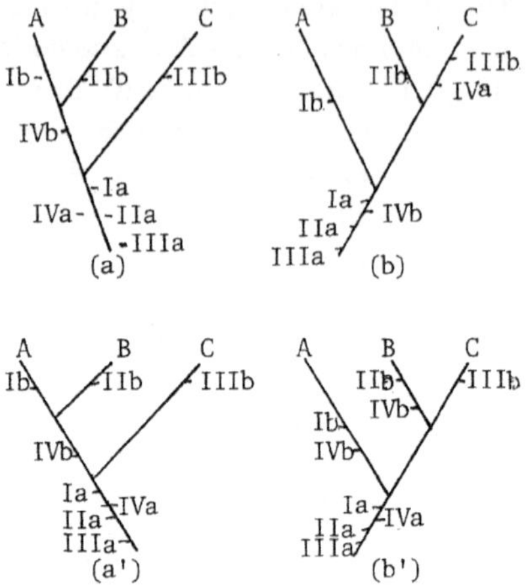

Fig. 1. The relationship between rooting with an outgroup and parsimony. See text for explanation.

the trees that can be hypothesized from the data are shown in figures 1a and 1b. There is no way to decide between these two trees on the basis of parsimony. In this situation, it appears that both require changes in four character states, so they are equally parsimonious. However, in the context of the higher level phylogeny, one of the trees is more so. The group outside the ABC taxa shows that tree 1b (see Figure 1a′) requires four character state changes, whereas tree 1b (see Figure 1b′) requires five such changes. Tree 1a is more parsimonious since it requires fewer character state changes. Therefore, the out-group comparison method allows us to determine the most parsimonious arrangement of taxa in the context of higher level phylogeny.

In protistology there are few published papers showing a cladistic analysis (but see Lynn, (1978) on colpodid ciliates, and Bremer and Wanntorp (1982) on plants). Lynn's analysis unfortunately uses common = primitive to polarize the character states. This method has been sharply criticized (Stevens, 1980; Watrous and Wheeler, 1981). Also the data matrix used by Lynn was not published, limiting the value of the work.

3. The Preferred Method of Classification

Having introduced the major approaches to classification, I shall now explain why I believe that cladistics best suits the needs of the protistologist.

3.1. Preference of Cladistics over Synthetics

Synthetic and cladistic taxonomy although both seek to reconstruct evolution, differ from each other in two fundamental respects: (1) the repeatability of the method, and (2) ranking taxa into categories.

Repeatability. Cladistics, since it follows a more precise analytical procedure for handling the data, is more repeatable and therefore more scientific. Synthetics is not repeatable and, therefore, not science but learned opinion.

A synthetic taxonomist must construct elaborate assumptions about the pathways of evolution. The synthetic classification is then based on these assumptions. Rarely do two synthetic taxonomists independently make the same assumptions; even if they do, they rarely construct the same classification (Kavanaugh, 1978). From this cause this method is not repeatable. Synthetics can also lead to circular reasoning if one attempts to use the classification to support any theory of the evolution of the group (Wagner, 1969).

Systematic hypotheses should be more than refined stories about evolutionary pathways. The cladistic method gives the taxonomist the ability to be more objective because it allows formulation of testable hypotheses about phylogenetic relationships among organisms without elaborate presupposition stories about those relationships. This is true because cladograms are formulated from an analysis of character data without previous evaluation or weighting of the data. The pathway of evolution is hypothesized from the classification without fear of circular reasoning. The hypothesis can be tested by cladistic analysis of new character data where agreement between the first and second sets of data strengthens the hypothesis and disagreement refutes it. Furthermore, such analyses are repeatable. A cladist can send his character data to any other cladist and both will construct exactly the same classification – even if the second taxonomist is unfamiliar with the taxa or the data.

Much work done by traditional synthetic taxonomists has been valuable, but the more formalized approach of cladistics is preferable. Many synthetic taxonomists appreciate this and commend cladistic methodology. Charig (1982), a synthetic paleontologist, states:

"... soon I began to realize that the more formal procedures of cladistic analysis were incomparably superior to the vague, slapdash, woolly methods used by us non-cladists. I read manuscripts for friends, I quickly became conscious of the fact that many of my non-cladistic colleagues – hard-working intelligent colleagues with high reputations – were collecting magnificent specimens in the field, preparing them beautifully, describing them very fully and with utmost clarity, and then writing accounts of their phylogenetic relationships that were nothing more than meaningless waffel."

Ranking Taxa. The second major difference between synthetic and cladistic methods is in the procedure of dividing up the phylogenetic tree into taxa and the subsequent ranking of those taxa into a hierarchy. The cladistic taxonomist believes that the main purpose of a classification of living organisms is to reflect their phylogenetic relationships (Hennig, 1966, p. 64). But the synthetic taxonomist maintains that sometimes it is desirable to remove one (or more) of the younger taxa from the other groups to which it is evolutionarily related and to rank it as high as the parent taxon (or even higher) if it has dramatically diverged from the older group.

This debate has largely degenerated into an argument over the definition of the terms monophyly, polyphyly, and paraphyly. The meaning of these words is surrounded by a great deal of confusion. The first two were introduced by Haeckel (1894) and the third by Hennig (1965). In recent years many authors have tried to discuss, redefine, or

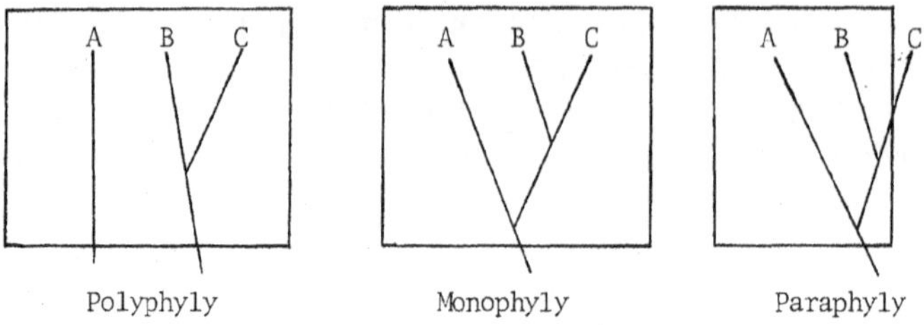

Fig. 2. Examples of polyphyletic, monophyletic, and paraphyletic groups.

clarify their meanings (e.g., see Ashlock, 1971, 1972, 1980; Bonde, 1977; Charig, 1976, 1982; Hennig, 1965, 1966; Nelson, 1971, 1973; Platnick, 1977, 1981; Simpson, 1961, 1971).

Essentially both types of taxonomists agree that a polyphyletic group is one that arises out of two or more ancestral taxa (Figure 2a); both identify polyphyletic groupings as artificial and arbitrary assemblages that should not be used in classification (Hennig, 1966; Simpson, 1961).

The confusion comes in the definition of monophyly. For many synthetic taxonomists, a group that originates from a single ancestral taxon and includes that ancestor within itself is said to be of monophyletic origin (Figure 2b and 2c). However, Hennig and all members of the cladistic school restrict the use of the word to those groups which include the ancestor and all the descendants (Figure 2b only). A group that excludes one or more of the descendant taxa is said to be paraphyletic (Figure 2c). Paraphyletic taxa may be acceptable to the synthetic but not to the cladistic taxonomist. Consequently, the cladistic classification more rigidly follows phylogeny. But the synthetic classification compromises phylogeny with degree of divergence.

For a common example used to clarify the difference (Mayr, 1974), substitute lizards for A, crocodiles for B, and birds for C in Figure 2c. The synthetic taxonomist would argue that a great deal of specialized evolution has occurred in the bird line, thus qualifying them for removal from the rest of the reptiles and given equal class status with them. On the other hand, a cladist would argue that this classification does not reflect the phylogeny of the group – crocodiles are more closely related to birds, not lizzards. Thus, the birds and crocodiles must be grouped together.

This example demonstrates the cladistic belief that taxa must necessarily correspond to natural groups (existing in reality outside the human mind) historically determined by the pattern of phylogeny. Syntheticists maintain that taxa may sometimes be entirely artificial concepts (existing only in the minds of their promulgators). The result is that the cladistic classification always attempts to reflect the phylogeny of the group. The synthetic classification does not.

3.2. Preference of Cladistics over Phenetics

Both cladistics and phenetics follow a prescribed procedure that makes their methods scientifically repeatable. Beyond this, however, they differ in several ways.

Reconstructing Evolution. Cladistics involves concepts and methods for estimating the branching patterns of evolutionary history. Phenetics illustrates only overall similarity. I maintain that, since taxonomic names are used to describe and discuss biological observations and theories, they should be consistent with the agency that has created the diversity of organisms (see also Farris, 1983b). Hennig (1966) and many other evolutionary biologists have pointed out that the relationships that lend cohesion to living and extinct organisms are genealogical relationships. All other types of relationships (e.g., phenotypic) are phenomena correlated with genealogical descent with modification (evolution!). It follows that the most descriptive, explanatory system for classifying the diversity of organisms will be one that seeks to recover the phylogeny of the organisms. Phenetics has entirely abandoned the idea of correlating classification with phylogeny and has thereby forfeited its ability to produce a useful general reference classification, the one to which we assign our Linnean nomenclature (Farris, 1983b). Although there are sometimes needs for specific purpose classifications to handle specific kinds of information and indicate particular relationships (Blackwelder, 1967), there is only one general reference classification. Naturally such a system will be more useful if it reflects the phylogeny of the taxa. Of course, pheneticists would argue that the desire to reflect phylogeny in classification merely reflects my taste and preference.

Information Content and Naturalness. Sneath and Sokal (1973) argue that the purpose of biological classification is not to reflect evolution, which is unknowable, but to create an indexing device through which storage and retrieval of information about organic diversity are facilitated. To fulfill this purpose they maintain that a classification should have high information content and be natural. They assert that grouping by overall similarity (phenetics) will best achieve such a classification . However, in a series of papers, Farris (1977, 1979, 1980, 1981a, 1981b, 1982, 1983a, 1983b) shows that cladistic methods will produce classifications that have more information content and naturalness than those resulting from the phenetic approach. Therefore, when judged by the phenetic criteria, cladistic classifications are superior to phenetic ones.

Information content was discussed and defined by Mill (1874, p. 466):

"The ends of scientific classification are best answered when objects are formed into groups respecting which greater number of general proposition more important, than could be made respecting any other groups into which the same things could be distributed."

Simply put, the information content of a classification is how much information it conveys about character state distribution. Information is contained in a classification when the branching pattern of the tree describes the distribution of the characters. For example, in Figure 2 are two trees for three taxa A, B, and C, and their common ancestor (CA) and a data chart for two characters. In tree –a, taxa A and B are closely aligned; C is distant. In tree –b, taxa B and C are closely aligned; A is distant. The characters change from state 0 to state 1 in A and B, indicating that A and B should be aligned. Consequently the branching pattern of tree –a describes the changes in the character data perfectly. Tree –b branches in a pattern that is opposite to the changes in the characters. Thus, tree –a has higher information content.

Characters	Taxa			
	A	B	C	Common Ancestor
I	1	1	0	0
II	1	1	0	0

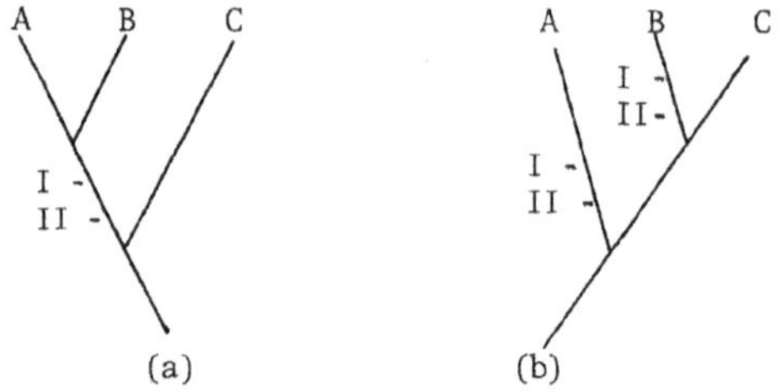

Fig. 3. An example of the parsimony criterion. Tree a is more parsimonious than tree b.

Information content can be indicated simply by counting the number of deviations of the character state distributions from the tree. The lower the number of deviations, the higher the information content (Schuh and Polhemus, 1981). In tree a, there are no deviations. In tree b there are two convergences of character states. State 1 in characters I and II appears to have arisen twice – once in taxon A and once in taxon B.

Farris (references cited above) has argued that grouping by synapormorphy (unique derived character states) rather than raw similarity (phenetics) will always give more informative classifications. This can be demonstrated by examining a phenogram and a cladogram of a group of ciliated protists for information content. The example (given below) uses data taken from a paper by Holt *et al.* (1974). I do not wish to argue for or against the way in which the authors of this paper have chosen characters, assigned character states, or decided which taxa to consider. Their data chart is used because (1) it is coded in a manner easily usable in cladistic and phenetic analysis, (2) it provides an example using 'real' protists, and (3) it is readily accessible in the published literature.

Table I shows data on five ciliates (from Holt *et al.*, 1974). Blanks in their data matrix on *Chaenea* were filled in by Riordan (1980, and personal communication). Figure 4 is a UWPGMA phenogram constructed using a simple matching coefficient (Sm). This procedure, which groups by overall similarity (see previous section on phenetic taxonomy), is the most commonly used phenetic method. Numerous books and articles explain this procedure in detail so it is unnecessary to do so here (e.g., see Sneath and Sokal, 1973, p. 230). Figure 5 shows a cladogram of the taxa given in Table I. The out-group used is *Actinobolina*, which has character state 0 for all the characters used. The cladogram was constructed following the Wagner method (for detailed discussion of this method see Farris, 1970, and Wiley 1982, p. 178).

TABLE I

Data from Holt *et al.* (1974)

	Lacrymaria	*Lagynophrya*	*Didinium*	*Chaenea*	*Dileptus*	*Actinobolina*
1. Nematodesmata of outer ring: (0) single; (1) paired	1	1	0	0	0	0
2. Fibrous annulus of outer ring: (0) not nd linked; (1) nd linked; (2) absent	0	0	1	2	2	0
3. Nematodesmata of inner ring: (0) unorganized; (1) hexagonal; (2) absent	0	0	0	2	1	0
4. Fibrous annulus of inner ring: (0) present; (1) absent	0	1	1	1	0	0
5. Cytopharynx: (0) permanent (1) not permanent	1	1	1	1	1	0
6. Cytostomial lips: (0) present; (1) absent	1	1	1	1	0	0
7. Cytostome: (0) apical (1) not apical	0	0	0	0	1	0
8. Corona: (0) alveolar (1) not alveolar	1	1	0	0	1	0
9. Corona: (0) subalveolar microtubules (1) no subalveolar microtubules	1	1	1	1	1	0
10. Corona: (0) transverse ribbons; no transverse ribbons	0	0	0	0	0	0
11. Coronal ciliature inner kinetosome: (0) ciliated; (1) non-ciliated	0	0	*	0	0	0
12. Coronal ciliature inner kinetosome: (0) transverse ribbon; (1) no trans.	0	0	*	0	0	0
13. Coronal ciliature inner kinetosome: (0) nematodesmata; (1) no nemato.	0	0	*	0	0	0
14. Coronal ciliature outer kinetosome: (0) ciliated; (1) not ciliated	0	0	*	0	0	0
15. Coronal ciliature outer kinetosome: (0) transverse ribbon; (1) no trans.	0	0	*	1	0	0
16. Coronal ciliature outer kinetosome: (0) postciliaries; (1) no postciliary	0	0	*	0	1	0
17. Coronal ciliature outer kinetosome: (0) linked to nd; (1) not linked	1	1	*	1	1	0
18. Outer-inner kinetosomal link: (0) absent; (1) present	0	1	*	1	1	0
19. Toxicysts: (0) 1 type; (1) 2 types; (2) 3 types	0	2	1	0	0	0
20. Toxicysts: (0) tentacle; (1) perioral; (2) intraoral	2	2	2	2	1	0
21. Mucocysts: (0) absent; (1) somatic; (3) somatic and coronal	2	1	2	2	2	0
22. Tela corticalis: (0) present; (1) absent	0	0	0	1	0	*
23. Clavate cilia: (0) present; (1) absent	0	0	0	*	*	0

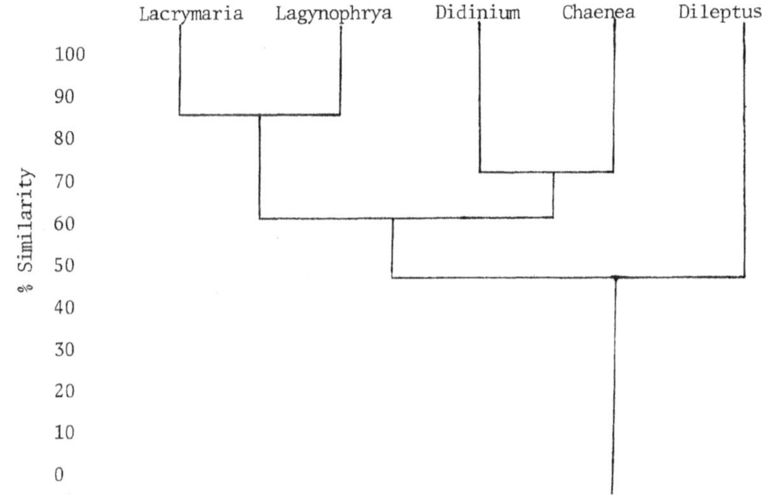

Fig. 4. Phenogram of the ciliates presented in Table I.

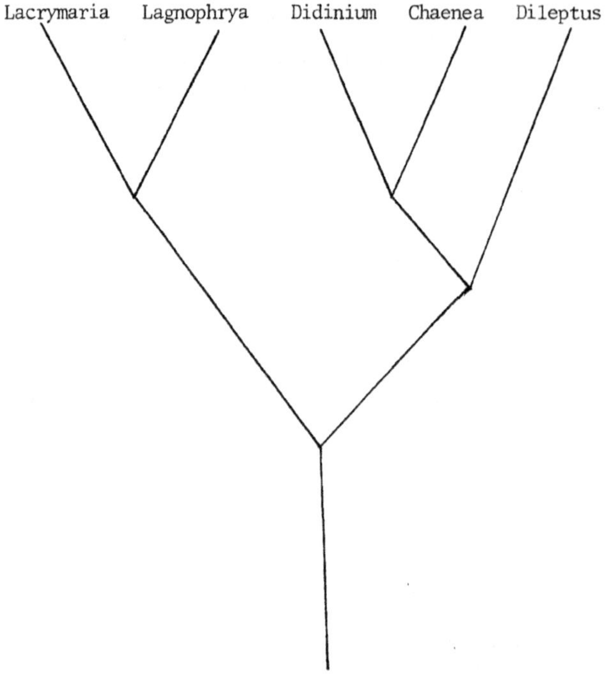

Fig. 5. Cladogram of the ciliates presented in Table I.

In the phenogram, five of the characters show convergence. These convergences occur in characters 2, 3, 4, 8, and 19. One of the characters (18) reverses. In the cladogram, only three of the characters show convergence and there are two reversals. Characters 5, 6, and 19 converge and characters 8 and 18 reverse. The branching pattern of the cladogram deviates from the character state distributions less than that shown by the phenogram. The cladogram, therefore, describes more of the character state changes and thus has a higher information content. Farris has demonstrated that this will always be the case because the cladistic method parsimoniously groups by advanced character states in a hierarchical branching sequence, whereas phenetics averages all the character data over the entire tree. There are two points to this argument.

First, the classification that allows the data to be described in the smallest number of character state changes corresponds to the most parsimonious tree. It is also the classification with the highest information content. Consider again Figure 3. In tree –a, character I changes from state 0 to state 1 only once. Character II also changes from state 0 to state 1 only once. Therefore, there are only two steps (character state changes) in tree –a. In tree –b, character I changes from 0 to 1 twice. Character II also changes from state 0 to 1 twice. There are four steps in tree –b. Because there are fewer steps in tree –a, it is the more parsimonious. It s obvious that trees that correspond best to the character state distributions have the fewest steps, that is they are the most parsimonious. Therefore, of any two classifications the one which is most parsimonious is the one with greater information content. Since cladistics seeks the most parsimonious tree and phenetics does not, it should always have higher (or at least equal if the trees are identical) information content than phenetics.

The second point is more complex and I direct readers to the papers of Farris (cited above) for a complete discussion. Briefly, taxonomists have long been aware that the branching pattern of classification will rarely reflect the distribution of all the character states. Usually, several characters corresponds closely but not exactly in the distribution of their states. The result is that a classification most informative about any one of those characters will differ slightly from those most informative about any other character. Pheneticists have generally maintained that this problem is best solved by averaging the data (overall similarity). This will create a classification in which every group is informative about each character but not in an exact way. If the distributions of the states of two characters differ, a compromised tree with a branching pattern intermediate between that dictated by the two characters is produced. The result of this is the loss of information content. A simple example will illustrate this point.

Figure 6 shows a data set of three characters for three taxa. Character I and character II indicate trees with different branching patterns. The phenogram calculated from these data (6a) shows a branching pattern intermediate between those indicated by the two characters. But the phenogram tells us only that A is different from B and both are different from C. The phenogram has no information content because it tells us nothing about the character state distributions. Furthermore, compromise is not necessary because biological classifications are hierarchical. Some groups in a hierarchy are subsets (or supersets) of other groups in the same classification. A simple consequence is that it is not possible for every group to be informative on every character considered

Characters	Taxa		
	A	B	C
I	0	1	1
II	0	0	1
III	0	1	0

Similarity			
	A	B	C
A	X		
B	33	X	
C	33	33	X

Fig. 6. Example of information content of a phenogram as compared to that of a cladogram. See text for explanation.

(Farris, 1980). For example, one character may describe relationships among the genera being classified; another character may describe the species of one of the genera. Unlike phenetics, the cladistic method describes the character state changes topographically on the tree because it joins taxa by actual shared character states. The taxa corresponding to characters that have different distributions are incorporated into a single classification, with the different distributions being placed at different branch points. Figure 6b shows the cladogram for the data in Figure 6. It can be seen from the cladogram that character II indicates the higher level relationship that B and C are more closely related than either are to A. Character I indicates that C is distinct from B. The cladogram has maximum information content – the character state changes are perfectly described.

It has been demonstrated by Farris (see citations above) that naturalness is equivalent to information content and that maximizing one will maximize the other. A natural classification, as defined by Whewell (1840, p. 521) is:

"... that arrangement obtained when one set of characters coincides with the arrangement obtained from another set."

The natural classification of a group is the one supported by the greatest agreement of the character state distributions. Since the classification with the highest information content will have the maximum congruence of characters, it will also be the most natural. Because cladograms have the highest information content, they will be the most natural.

3.3. Complaints against the cladistic method

Cladistic taxonomy has been criticized because it fails to provide an empirical method for choosing a monophyletic taxon to be classified (and by default, choosing what is the out-group) for distinguishing homologous character characters from convergent characters, and for assigning character states. First, it must be pointed out that all taxonomists must choose taxa and characters and assign character states. Therefore, it is inappropriate to criticize only the cladistic (or any other) method for a general problem that all taxonomic methods face. These problems are not ignored by cladists. Many cladistic taxonomists (e.g., see textbooks such as Eldredge and Cracraft, 1980; Nelson and Platnick, 1981; Wiley, 1981) have offered guidelines for determining in-groups and out-groups. When no out-group can be determined, Farris (1972) has suggested rooting the tree at a point between the two taxa that shares the fewest character states (mid-point rooting). The advantages and disadvantages of mid-point rooting are discussed by Farris.

Cladistic methods do allow taxonomists to test their hypotheses of monophyly or homology by corroborating their initial results using higher level taxa or additional character data. Phenetic and synthetic methods cannot always be used in the same way (see Mickevitch, 1978). As stated in the introduction, it is not my intention to explore this aspect of taxonomy in this paper (see discussion of the topic in Wiley, 1981).

4. Conclusions

The superiority of cladistic methods over both synthetic and phenetic methods is briefly advanced and reviewed above. Cladistics creates testable scientific hypotheses of phylogeny that also give the most informative summary of available data. Thus it best fits the criteria for a method for determining the general reference classification in biology.

For protistologists in particular, cladistics is especially useful. Inundated by a mind-boggling abundance of ultrastructural, biochemical, and cell biological information, protistologists could be greatly helped by the highly informative way in which cladistics orders and summarizes the data. In addition to classifying protist taxa, hypotheses about the evolution of cell organelles and cellular processes could be scientifically formulated and tested by cladistics. Because cladistic classifications best summarize the data, they would also be best for making predictions about taxa and characters (Farris, 1980). They would, for the same reason, be the most stable (Farris, 1980). This is a welcome thought for protistology, where classifications sometimes seem to change as fast as scientific journals appear. It is for all of these reasons that cladistic methods best suit the needs of the evolutionary protistologist.

References

Ashlock, P. D.: 1971, *Syst. Zool.* **21**, 63–69.

Ashlock, P. D.: 1972, *Syst. Zool.* **21** (year 1971), 430–438.

Ashlock, P. D.: 1980, *Syst. Zool.* **28** (year 1979), 441–450.

Blackwelder, R. E.: 1967, *Taxonomy: a Text and Reference Book*, John Wiley and Sons, New York.

Bonde, N.: 1977, Cladistic Classification as Applied to Vertebrates, in M. K. Hecht, P. C. Goody, and B. M. Hecht (eds.), *Major Patterns in Vertebrate Evolution*, Plenum Press, New York.

Bremer, J. and Wanntorp, L.: 1982, in V. A. Funk and D. R. Brooks (eds.), *Advances in Cladistics*, New York Botanical Garden, New York.

Charig, A. J.: 1976, Dinosaur Monophyly and a New Class of Vertebrates, in A. A. Bellaris and C. B. Cox (eds.), *Morphology and Biology of Reptiles*, Academic Press, New York.

Charig, A. J.: 1982, Systematics in Biology: A Fundamental Comparison of Some Major Schools of Thought, in K. A. Joysey and A. E. Friday (eds.), *Problems of Phylogenetic Reconstruction*, Academic Press, New York.

Copeleston, F. C.: 1972, *A History of Medieval Philosophy*, Methuen, London.

Crowson, R. A.: 1982, Computers Versus Imagination in the Reconstruction of Phylogeny, in K. A. Joysey and A. E. Friday (eds.), *Problems of Phylogenetic Reconstruction*, Academic Press, New York.

Eldredge, N. and Cracraft, J.: 1980, *Phylogenetic Patterns and the Evolutionary Process*, Columbia University Press, New York.

Farris, J. S.: 1970, *Syst. Zool.* **18**, 279–285.

Farris, J. S.: 1972, *Amer. Nat.* **106**, 645–668.

Farris, J. S.: 1977, On the Phenetic Approach to Vertebrate Classification, in M. K. Hecht, P. C. Goofy, and B. M. Hecht (eds.), *Major Patterns in Vertebrate, Evolution*, Plenum Press, New York.

Farris, J. S.: 1979, *Syst. Zool.* **28**, 200–214.

Farris, J. S.: 1980, *Syst. Zool.* **28**(year 1979), 483–519.

Farris, J. S.: 1981a, *Syst. Zool.* **29**(year 1980), 360–381.

Farris, J. S.: 1981b, *Syst. Zool.* **29**(year 1980), 386–401.

Farris, J. S.: 1982, *Syst. Zool.* **31**, 314–320.

Farris, J. S.: 1983a, *Syst. Zool.* **31**(year 1982), 413–444.

Farris, J. S.: 1983b, The Logical Basis of Phylogenetic Analysis, in V. A. Funk and N. I. Platnick (eds.), *Advances in Cladistics II*. Columbia University Press, New York.
Friz, C. T.: 1979, *Arch. Protistenk.* **122**, 360–371.
Haeckel, E.: 1894, *Systematische Phylogenie. I. Systematische Phylogenie der Protisten und Planzen*, Reimer, Berlin.
Hennig, W.: 1950, *Grundzuge einer Theorie der phylogenetischen Systematik*, Deutsche Zentralverlag, Berlin.
Hennig, W.: 1965, *Ann. Rev. Ent.* **10**, 97–110.
Hennig, W.: 1966, *Phylogenetic Systematics*, University of Illinois Press, Urbana.
Holt, P. A., Lynn, D. H., and Corliss, J. O.: 1974, *Protistologica* **9**, 521–541.
Kavanaugh, D. H.: 1978, Hennigian Phylogenetics in Contemporary Systematics: Principles, Methods and Uses, in J. S. Romberger (ed.), *Biosystematics in Agriculture*, John Wiley and Sons, New York.
Kneale, W. and Kneale, M.: 1962, *The Development of Logic*, Clarendon Press, Oxford.
Lundberg, J. G.: 1972, *Syst. Zool.* **21**, 398–413.
Lynn, D. H.: 1978, *BioSystems* **10**, 201–211.
Lynn, D. H.: 1979a, *Trans. Amer. Micros. Soc.* **98**, 353–368.
Lynn, D. H.: 1979b, *J. Protozool.* **26**, 359–366.
Mayr, E.: 1969, *Principles of Systematic Zoology*, McGraw-Hill, New York.
Mayr, E.: 1974, *Z. Zool. Syst. EvolForsch.* **12**, 94–312.
Mickevitch, M.: 1978, *Syst. Zool.* **27**, 143–158.
Mill, J. S.: 1874, *A System of Logic, Ratiocinative and Inductive, being a Connected View of the Principles of Evidence and the Methods of Scientific Investigation*, 8th ed., Longsman, Green and Co., London.
Nelson, G. J.: 1971, *Syst. Zool.* **20**, 471–472.
Nelson, G. J.: 1973, *Syst. Zool.* **22**, 310–312.
Nelson, G. J. and Platnick, N. I.: 1981, *Systematics and Biogeography*, Columbia University Press, New York.
Platnick, N. I.: 1977, *Syst. Zool.* **26**, 195–200.
Platnick, N. I.: 1981, *Syst. Zool.* **29**(year 1980), 537–546.
Riordan, G. P.: 1980, *J. Protozool.* **27**, p. 22A.
Russell, B.: 1946, *History of Western Philosophy*, George Allen and Unwin, London.
Schuh, R. T. and Polhemus, J. T.: 1980, *Syst. Zool.* **29**, 1–26.
Simpson, G. G.: 1961, *Principles of Animal Taxonomy*, Columbia University Press, New York.
Simpson, G. G.: 1971, Concluding Remarks: Mesozoic Mammals Revisited, in D. M. Kermack and K. A. Kermack (eds.), *Early Mammals*, Academic Press, New York.
Small, E. B. and Lynn, D. H.: 1981, *BioSystems* **14**, 387–401.
Sneath, P. H. A. and Sokal, R. R.: 1973, *Numerical Taxonomy*, W. H. Freeman and Co., San Francisco.
Steven, P. D.: 1980, *Ann. Rev. Ecol. Syst.* **11**, 333–358.
Van Valkenburg, S. D., Karlander, E. P., Patterson, G. W., and Colwell, R. R.: 1975, *Taxon* **26**, 497–505.
Wagner, H.: 1969, The Construction of a Classification, in Sibley, C. G. (ed.), *Proceedings of an International Conference Conducted at the University of Michigan, Ann Arbor*. National Academy of Science, Washington, D.C.
Watrous, L. E. and Wheeler, Q. D.: 1981, *Syst. Zool.* **30**, 1–11.
Whewell, W.: 1840, *The Philosophy of Inductive Sciences, Founded upon Their History*, Parker, London.
Wiley, E. O.: 1981, *Phylogenetics: the Theory and Practice of Phylogenetic Systematics*, John Wiley and Sons, New York.

A BIOMETRICAL ANALYSIS OF MORPHOLOGICAL VARIATION WITHIN A SECTION OF GENUS *EUPLOTES* (CILIATA, HYPOTRICHIDA), WITH SPECIAL REFERENCE TO THE *E. VANNUS* COMPLEX OF SIBLING SPECIES

VÉRONIQUE MACHELON and JEAN GÉNERMONT

Laboratoire de Zoologie 2, Université de Paris-Sud, 91405 Orsay Cedex, France

and

YVETTE DATTÉE

Laboratoire d'Amélioration des Plantes, Université de Paris-Sud, 91405 Orsay Cedex, France

(Received 12 December, 1983)

Abstract. Some 50 clones belonging to a rather homogeneous, presumably monophyletic, section of genus *Euplotes* proved to be distributed between two clear-cut subgroups on the basis of cell size and relative peristome length. These subgroups are referred to as *vannus* and *minuta* morphotypes.

Within *vannus* morphotype, the *E. vannus* complex of 5 sibling species was previously defined on the basis of interfertility criterion. Electrophoretical studies revealed clear-cut biochemical differences between the *E. vannus* complex and the *minuta* morphotype.

A multivariate analysis concerning 6 characters shows that the 5 species in the complex are closely similar. Nevertheless, the results of discriminant analysis suggest that a morphological discrimination could be possible.

The status of one peculiar clone remains somewhat obscure. It belongs undoubtedly to the *vannus* morphotype, but shows no mating reaction with any of the 5 species of *E. vannus* complex and is electrophoretically related to *minuta* morphotype.

The following interpretation is suggested. In this section of genus *Euplotes*, two stable adaptive zones are available, corresponding to the two morphotypes. Speciation and change of adaptive zone are more or less independent evolutionary events. For example, the 5 species of the *vannus* complex have been generated by speciation without change of adaptive zone. Conversely, the exceptional clone may be phyletically related to the *minuta* morphotype through a rather recent jump from one adaptive zone to the other.

1. Introduction

Among protozoa, the use of the so-called biological concept of species is practically restricted to a few genera of ciliates. It is thus known that some taxa which had been for a long time thought to be good species are actually made up of a lot of strictly intersterile, though closely similar at the morphological level, species (see by review Génermont, 1976). The best examples of such a situation are the two complexes '*Paramecium aurelia*' and '*Tetrahymena pyriformis*': each of the component species is now provided with a proper binomen (Sonneborn, 1975; Nanney and MacCoy, 1976; Nyberg, 1981). These two complexes belong to the order Hymenostomatida, a taxon in which slight morphological differences are hardly detectable, since the cell bears a very large number of cilia, quasi-uniformly distributed on the whole surface. In fact, the discrimination between the species of the *P. aurelia* seems to be impossible, even by

Origins of Life **13** (1984) 249-267. 0302-1688/84/0133-0249$02.85.

means of multivariate methods (Gates *et al.*, 1974; Gates and Berger, 1976).

In Hypotrichs, the morphological patterns can be acurately described, since the cell exhibits a highly differentiated cortical organization. It is thus expected that clear-cut distinctive features are to be more easily found when closely related species are compared. The present work is an attempt to identify reliable morphological characters suitable for species discrimination within a section of genus *Euplotes*.

According to Curds (1975), the section *A* of this genus includes 6 closely related species, the definition of which is founded on (i) the number of dorsal ciliary rows and (ii) the number of cilia in central rows. Four of them were previously listed by Tuffrau (1960) who put forward some additional discriminating features, such as cell size, macronuclear shape and number of caudal cirri. It is now evident from studies on morphology and morphogenesis as well as from breeding analyses (Frankel, 1973; Génermont *et al.*, 1976; Gates, 1977; Laloe, 1979) that this conventional taxonomy is obsolete. On the basis of interfertility criterion, five intersterile species have been characterized, making up together the "*Euplotes vannus*" complex (Génermont, 1976; Génermont *et al.*, 1976; Hallet, 1978; Machelon, 1978, 1982). This complex includes the conventional taxa *E. vannus* Müller 1786, *E. crassus* Dujardin 1841 and *E. mutabilis* Tuffrau 1960.

The present contribution is a part of a more comprehensive study, the aim of which is to elucidate evolutionary processes within the complex (affinities between species, isolating mechanisms, modes of speciation, patterns of morphological, biochemical and physiological differentiation, etc.) and relationships of the complex with other *Euplotes*. Some of the results have been published elsewhere: those dealing with isolating mechanisms (Machelon, 1982) and biochemical variation (Machelon and Demar-Gervais, 1983).

2. Material and Methods

2.1. Collection and maintenance of clones

Several dozen clones have been isolated from samples of sea-water collected in different countries (France, Italy, Tunisia, Algeria, Morocco, U.S.A., West Indies). They are cultivated in monoxenic conditions (dried lettuce infusion prepared with sea-water, sterilized, then inoculated with *Enterobacter aerogenes*) at room temperature.

2.2. Morphometry

The samples were extracted from cultures grown up to the beginning of stationary phase. This procedure precludes any morphogenetic activity (binary fission, sexual phenomena, physiological reorganization) and reduces as far as possible environmental variability from one sample to another. The samples are silver-impregnated according to Chatton and Lwoff (1930), with the modifications proposed by Frankel and Heckmann (1968).

The following measurements have been taken on individual cells (Figure 1):

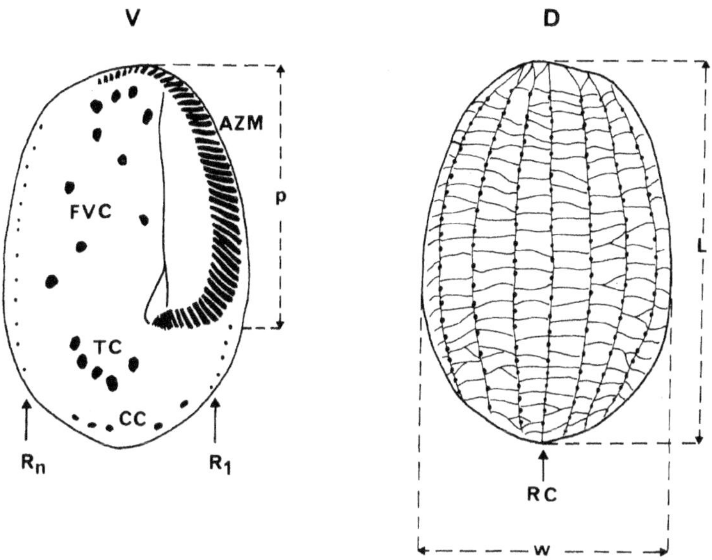

Fig. 1. Main morphological features analizable in *Euplotes*: V, ventral view; D, dorsal view; AZM, adoral zone of membranelles; FVC, frontoventral cirri; CC, caudal cirri; TC, transverse cirri; R_1, dorsal ciliary row no. 1; R_n, dorsal ciliary row no. n; RC, central dorsal ciliary row; L, cell length; w, cell width; p, peristome length.

– cell length, L;
– maximal cell width, w;
– peristome length, i.e. distance extending from the anterior end of the cell to the mouth, p;
– number of dorsal ciliary rows, R;
– number of cilia in the central row (if the n rows are numbered from left to right the central one is either no. $(n + 1)/2$ or no. $n/2$ according that n is odd or even; in fact kinesotomes rather than cilia are counted), C;
– number of caudal cirri, Q.

The first three variables are related to cell size and shape. In order to avoid intrinsic correlations (examplified in Figure 2), two additional variables were introduced: $W = w/L$ (relative cell width) and $P = p/L$ (relative peristome length); they were used in most analyses instead of w and p; they are representative of cell shape, while L is representative of cell size. The last three variables reflect the cortical organization; they are of current use in conventional taxonomy.

Many characters *a priori* available for morphometric analysis were not taken into consideration for various reasons. The number of transverse cirri is quite stable throughout the genus; the number of fronto-ventral ones is stable within the section under study. The number of membranelles is hardly estimated, since the consecutive membranelles are frequently poorly separated in the posterior part of the adoral zone. Irregularities of dorsal argyrome (*mutabilis* phenotype) are not expected to be observed

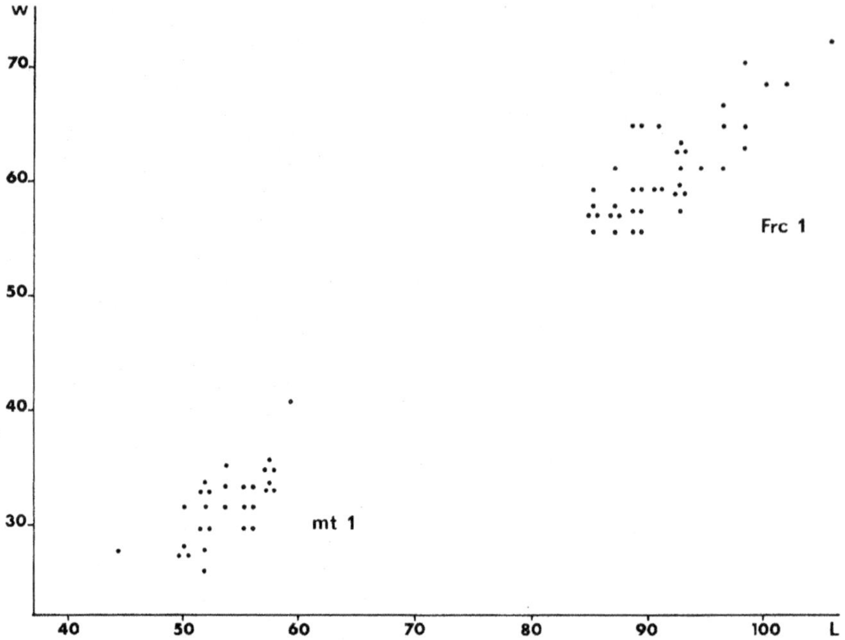

Fig. 2. Relation between cell length (L) and cell width (w) in two clones, the largest (*Frc 1*) and the smallest (*mt 1*) of the sample. Within each clone, a very strong positive correlation is observed between L and w, such as in the two clones taken as examples. At the interclonal level, a similar correlation exists between the clonal means, $\overline{L}$ and $\overline{w}$. The correlation coefficients are estimated as 0.737 in the *vannus* morphotype and 0.857 in *minuta*: these estimates are not significantly different. The linear relation between $\overline{L}$ and $\overline{w}$ is well described by the major axes of the dispersion diagrams, the equations of which are:

$\overline{w} = 1.123\,\overline{L} - 37.576$ for *vannus*

and

$\overline{w} = 1.068\,\overline{L} - 21.755$ for *minuta*

These two lines are approximately parallel (their slopes are not significantly different).

during stationary phase of growth; furthermore, their quantification is rather problematic. As for macronuclear shape, its variation seems strongly dependent on physiological conditions which are not subject to experimental control.

Unfortunately, for technical reasons, the 6 measurements are not available for each individual cell. Consequently, it was not possible to analyse intraclonally the joint distributions of two (or more) variables. This could be done only interclonally on the basis of clonal means. In each clone, the mean $\overline{X}$ of the variable X is estimated from at least 20 individual measurements (except for some exceptional clones).

2.3. Mathematical analyses of quantitative data

Since different mathematical procedures have been used in order to solve different problems, relevant informations about them will be given in the corresponding sections of the chapter 'Results'.

3. Results

3.1. Separation of *vannus* and *minuta* morphotypes on the basis of cell length

According to Curds (1975) the range of variation for cell length in *Euplotes*, section *A*, is rather large: from *E. minuta* (54 μm) to *E. crassus* (100–130 μm). A first task was to analyse this heterogeneity, in order to detect eventual discontinuities. The sample under study was made up of 34 clones, including 1380 cells. The results are listed in

TABLE I

Variation of cell length (μm) among a sample of 34 clones. Upper part of the table: *vannus* morphotype; lower: *minuta* morphotype. Specific ascertainment is stated for those clones which belong to *E. vannus* complex.

Clone	Species	Sample size	Mean	Standard deviation
Frc 1	2	65	91.42	5.27
Mi 1.10	5	23	91.14	6.95
Da 17	4	26	90.28	5.90
Ay 1.1	4	50	89.22	5.94
Cn 1.1	3	41	88.80	5.86
Bou 2.3	2	60	88.12	5.35
PP 4	5	24	87.65	6.38
Ro	2	50	87.52	5.09
Bou 1	1	23	85.11	6.03
Ay 1.4	4	62	85.10	5.45
Bou 2.4	4	47	84.67	6.54
Cr 1.7	3	29	84.04	5.86
Bou 2.1	2	36	83.90	4.76
Ba 2.3	1	26	83.05	8.11
Arc 1	4	24	82.64	5.20
CC 1.6	5	26	82.48	4.69
Al 12	1	25	81.78	5.99
Cn 2.2	3	42	81.53	6.11
Ré 1.9	1	23	81.48	5.61
Da 4	3	51	81.41	5.25
Da 2	3	54	80.21	7.33
Fm 1.2	3	25	80.00	6.46
Frc 3		44	79.37	7.33
2 d	2	49	77.78	6.83
Sa	2	25	73.11	5.32
At 1		48	65.89	6.93
Frd		38	61.40	5.59
Ag 5		25	58.96	6.29
Ag 4		43	58.10	5.31
Mic 1		85	57.54	6.78
Ag 1		17	57.52	5.38
Lo		49	55.56	4.34
Aq 1		21	54.94	5.35
mt 1		30	53.89	3.38

Table I, in which the clones are ordinated according to decreasing values of means. These means have been compared two by two, following the so-called 'corrected-*t*' method. When two consecutive means are compared, the difference is never significant at 0.05 level (total α-risk) except for clones *Sa* and *At 1*. More precisely, the whole set of clones may be divided into two groups, in such a way that any mean of one group is significantly different from any mean of the other group (at $\simeq 10^{-6}$ level), while no discontinuity is detected within each group (at 0.05 level).

In the first group, the values of $\bar{L}$ (mean cell length) range approximately from 73 to 92 μm; in the second one, from 53 to 66. On the basis of conventional taxonomy, these ranges do not differ from those which are expected for two discrete classes, the former including the four taxa *E. vannus* Müller 1786, *E. crassus* Dujardin 1841, *E. balticus* Kahl 1932 and *E. mutabilis* Tuffrau 1960, and the latter including *E. minuta* Yocum 1930 and *E. cristatus* Kahl 1932. In the present paper, these two groups will be referred to as the '*vannus* morphotype' and the '*minuta* morphotype', respectively.

At this point it must be recalled that in previous studies five species, numbered from 1 to 5, have been characterized on the basis of interfertility criterion, making together the '*E. vannus* complex' (see Machelon, 1982). On the same basis, the specific identity has been undoubtedly established for 24 out of the 25 clones belonging to the *vannus* morphotype. The last one, *Frc 3* (which is rather short, but not the shortest) could not be classified, since it displayed no sexual reaction with any of the standard clones of the 5 species; moreover, the biochemical data seem to indicate that it is more closely related to the *minuta* morphotype than to the *vannus* one.

3.2. Comparison of the two morphotypes on the basis of cell shape

For each of the two variables accounting for cell shape, i.e. *W* (relative cell width) and *P* (relative peristome length), the comparison of the two morphotypes was achieved by means of hierarchical analysis of variance, according to the following model:

$$X_{ijk} = \bar{X} + A_i + B_{ij} + C_{ijk}$$

with:

X_{ijk}: value observed on the *k*th cell in the *j*th clone belonging to the *i*th morphotype,
$\bar{X}$: general mean,
A_i: effect of the *i*th morphotype,
B_{ij}: effect of the *j*th clone within the *i*th morphotype,
C_{ijk}: individual deviate.

The results of these analyses are summarized in Tables II and III.

As for relative width, the only significant effect is that of clone, while the two morphotypes appear as closely similar. It is thus likely that the overall shape of the cell, irrespective of size, is genetically controlled, resulting in parallel variation patterns within the two morphotypes. Conversely, the effect of morphotype on relative peristome length is highly significant. This probably indicates that this attribute is subject to some size-dependent constraints. Actually, *P* is smaller in *minuta* than in

TABLE II

Hierarchical analysis of variance of relative width (whole set of data)

Variation	Degrees of freedom	Sum of squares	Mean square	F
Total	998	3.678		
Between morphotypes	1	0.0570	0.0570	$1.3^{\circ\circ}$
Between clones within morphotype	33	1.443	0.0437	$19.4^{\circ\circ}$
Between cells within clone	964	2.178	0.00226	

TABLE III

Hierarchical analysis of variance of relative peristome length (whole set of data)

Variation	Degrees of freedom	Sum of squares	Mean square	F
Total	747	1.995		
Between morphotypes	1	0.3309	0.3309	$29.5^{\circ\circ}$
Between clones within morphotype	31	0.3482	0.0112	$6.1^{\circ\circ}$
Between cells within clone	715	1.316	0.00184	

vannus morphotype. This conformational difference might be related to some biological specialization in natural conditions: *minuta* probably feeds mainly on bacteria or very small protists, while *vannus* is able to ingest much larger prey.

It ensues that the discrimination between the two morphotypes is more efficient on the basis of the joint distribution of $\overline{L}$ and $\overline{P}$ than on the basis of $\overline{L}$ alone (Figure 3). An outstanding feature of this distribution is the negative correlation between the two variables within each morphotype: the correlation coefficient is estimated as -0.387 for *vannus* and -0.608 for *minuta* (weighted average estimate: -0.432, different from zero at 0.02 level). This indicates that within the range of variation of each morphotype, the absolute peristome length is more stable than cell length, a pattern which might be related to the stability of feeding habits within a morphotype.

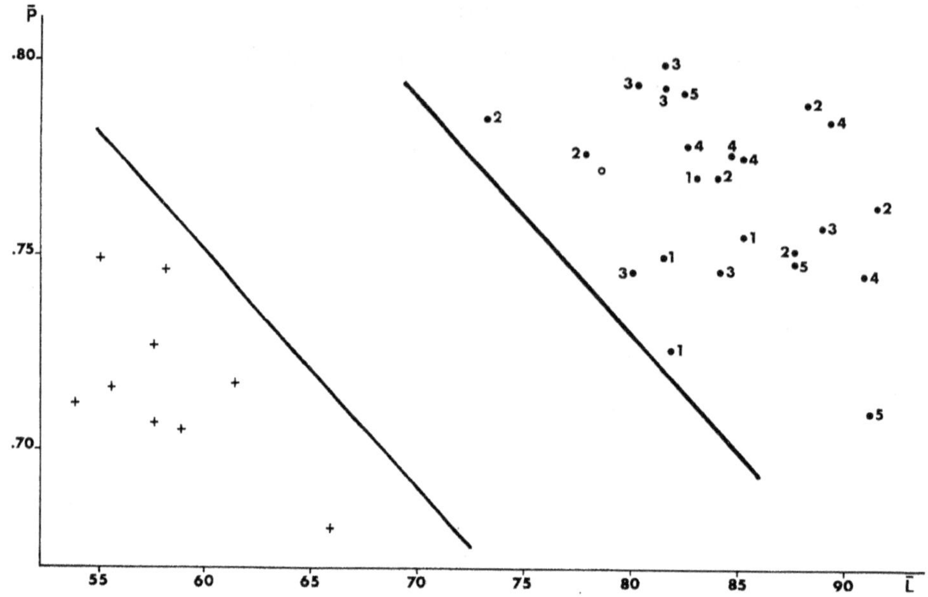

Fig. 3. Relation between mean cell lenght ($\bar{L}$) and mean relative peristome length ($\bar{P}$), with graphs of the two linear relations, $\bar{P} + 0.006\bar{L} = 1.11$ and $\bar{P} + 0.006\bar{L} = 1.21$. These two lines have been empirically determined in such a way that (*i*) they limit the 'no man's land' between the areas of the *minuta* (at left) and *vannus* (at right) morphotypes; (*ii*) they are parallel to each other; (*iii*) they are roughly parallel to the major axes of the two sets of points. Circles: *vannus* morphotype (full: *vannus* complex; open: *Frc 3*); crosses: *minuta* morphotype; the numbers refer to species within the *E. vannus* complex.

3.3. Variation of cell size and shape among the *E. vannus* complex

The variation of *L*, *W*, and *P* within the *E. vannus* complex has been studied by means of univariate analysis of variance, following the same model as above, except that *i* refers to the species which the clone belongs to. The results are summarized in Tables IV to VI. The three analyses yielded similar results: no significant heterogeneity between species, highly significant heterogeneity between clones within species. Thus, while some genetic variability for cell size and shape occurs within each species, the level of evolutionary divergence for these characters between species remains very low.

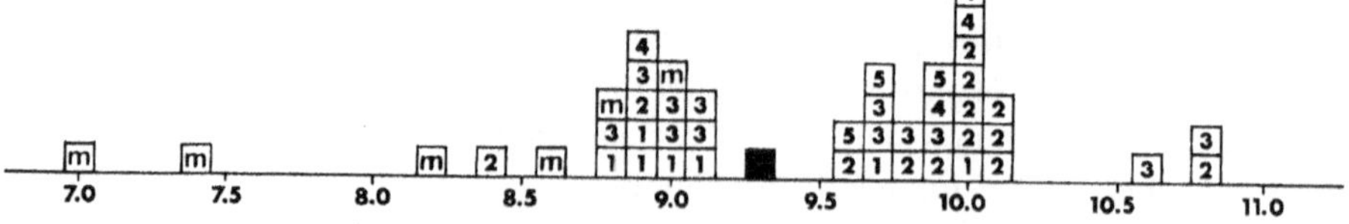

Fig. 4. Distribution of 47 clones according to their mean number of dorsal ciliary rows. Each clone is represented by a square; the letter *m* refers to the *minuta* morphotype; the numbers refer to the species within *E. vannus* complex; the full square is representative of clone *Frc 3*.

TABLE IV

Hierarchical analysis of variance of cell length (*E. vannus* complex)

Variation	Degrees of freedom	Sum of squares	Mean square	F
Total	905	14.052		
Between species	4	0.715	178.8	< 1
Between clones within species	19	4.250	223.6	$21.7^{\circ\circ}$
Between cells within clone	882	9.087	10.3	

TABLE V

Hierarchical analysis of variance of relative cell width (*E. vannus* complex)

Variation	Degrees of freedom	Sum of squares	Mean square	F
Total	684	2.203		
Between species	4	0.0634	0.0159	< 1
Between clones within species	19	0.7032	0.0370	$17.0^{\circ\circ}$
Between cells within clone	661	1.437	0.00217	

TABLE VI

Hierarchical analysis of variance of relative peristome length (*E. vannus* complex)

Variation	Degrees of freedom	Sum of squares	Mean square	F
Total	564	1.157		
Between species	4	0.0511	0.0128	1.1
Between clones within species	19	0.2222	0.0117	$7.2^{\circ\circ}$
Between cells within clone	541	0.8840	0.00163	

3.4. Variation of cortical patterns

Since the number of dorsal ciliary rows has been for many years accepted as the most powerful criterion suitable for species discrimination (Tuffrau, 1960; Curds, 1975), this character deserves special attention.

The distribution of the mean number of ciliary rows among a sample of 47 clones exhibiting either *vannus* (41 clones) or *minuta* (6 clones) morphotype is illustrated in Figure 4. The most prominent feature of this distribution is its plurimodality, with two main modes close to 9.0 and 10.0. This is in good agreement with previous conclusions of Frankel (1975): in a given clone, the number of rows is not fully stable, but one of the possible phenotypes is more stable than the others, this peculiar phenotype being referred to as the stability center. In the sample under study, most clones show either 10 or 9 rows as stability centers. For three clones, the stable number is probably 11, and the transition from phenotype 11 to phenotype 12 is probably exceptional, while transitions from 11 to 10 and from 10 to 11 do occur in the course of asexual growth.

Obviously, the number of rows is generally lower in *minuta* than in *vannus* morphotype. Nevertheless the two ranges overlap at the level of stability center 9. Within the *E. vannus* complex, the mean number of dorsal ciliary rows do not allow the specific ascertainment of a clone. Moreover, interspecific variability does not significantly exceed intraspecific one (Kruskal-Wallis test).

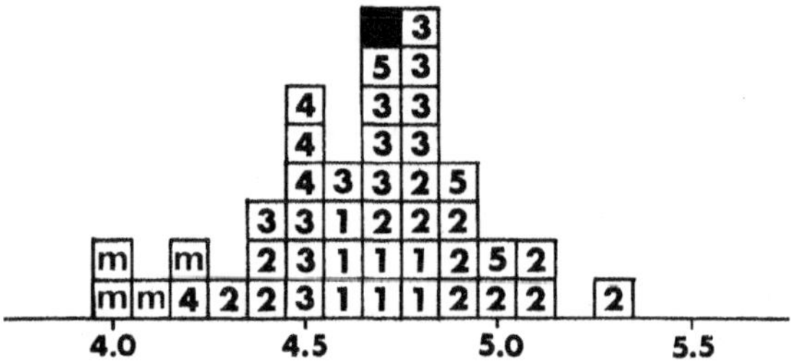

Fig. 5. Distribution of 44 clones according to their mean number of caudal cirri (same symbols as in Figure 4).

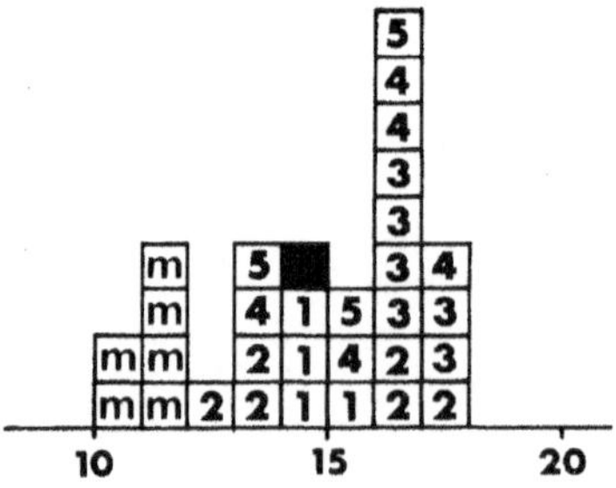

Fig. 6. Distribution of 31 clones according to their mean number of cilia in central dorsal row (same symbols as in Figure 4).

The variation patterns of the two other cortical variables are illustrated in Figures 5 and 6. They are obviously of very poor taxonomic value at least for discrimination of species within *E. vannus* complex.

From morphological and morphogenetical point of views, some correlations are expected between cortical variables and metric ones: for instance, positive correlations between cell length and number of cilia in central row, on one hand, and between cell width and number of ciliary rows, on the other hand.

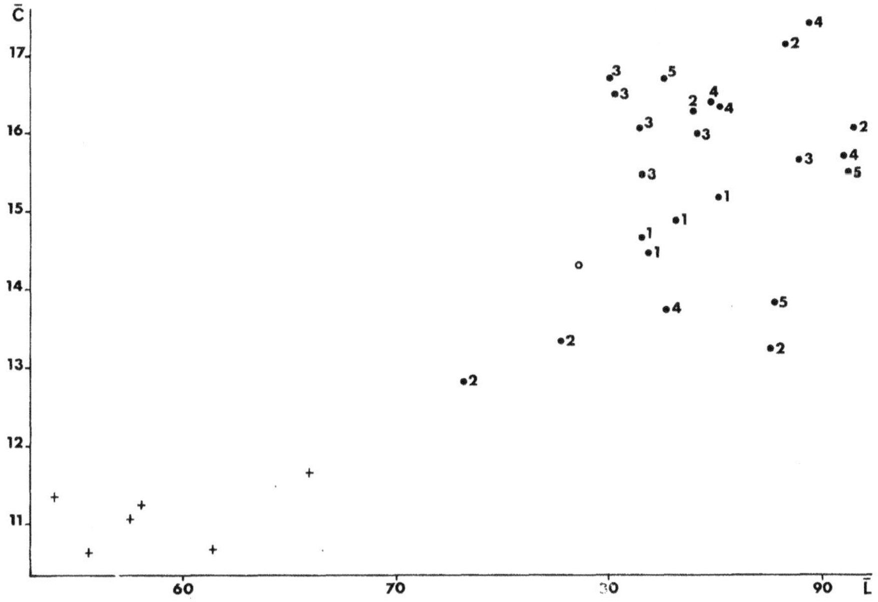

Fig. 7. Relation between mean cell length ($\bar{L}$) and mean number of cilia in central dorsal row ($\bar{C}$) in a set of 31 clones (same symbols as in Figure 3).

A positive correlation is actually observed between $\bar{L}$ and $\bar{C}$ in a sample of 31 clones (*vannus* morphotype: 25 clones; *minuta*: 6) (Figure 7). At the level of the whole sample, this result could be inferred from the results of univariate analyses, since it has been shown that both $\bar{L}$ and $\bar{C}$ are lower in *minuta* than in *vannus*. At the level of each morphotype, the estimates of the correlation coefficients are 0.389 for *vannus* and 0.107 for *minuta*; since they are not significantly different from each other, a combined estimate of intramorphotype correlation coefficient has been computed: 0.359, a value which is just significantly higher than zero at 0.05 level (one-tailed test).

The relation between $\bar{w}$ and $\bar{R}$ has been studied by means of non-parametric methods, since the distribution of $\bar{R}$ is very far from normal (see Figure 4). The Kendall rank correlation coefficient, estimated from a sample of 30 clones (*vannus*: 24 clones; *minuta*: 6), is $\tau_{\bar{w}\bar{R}} = 0.424$, which differs significantly from zero at 0.01 level. However, this rather strong correlation might proceed from an association between each of the two variables and cell size. Thus we have computed the partial rank correlation coefficient between $\overline{w}$ and $\overline{R}$ independently of variation of $\overline{L}$. The estimate is much lower than the previous one, though still positive: $\tau_{\overline{w}\overline{R}\,\cdot\,\overline{L}} = 0.179$. Unfortunately, as far as we know, no method is available for testing the significance of such an estimate.

3.5. Survey of variation by means of principal components analysis

Each of 26 clones (*vannus*: 21; *minuta*: 5) has been characterized by the means of the 6 variables L, W, P, R, C, and Q. The corresponding set of data has been studied by means of principal components analysis, on the basis of the correlation matrix (rather than the variance-covariance one, owing to the heterogeneity of variables).

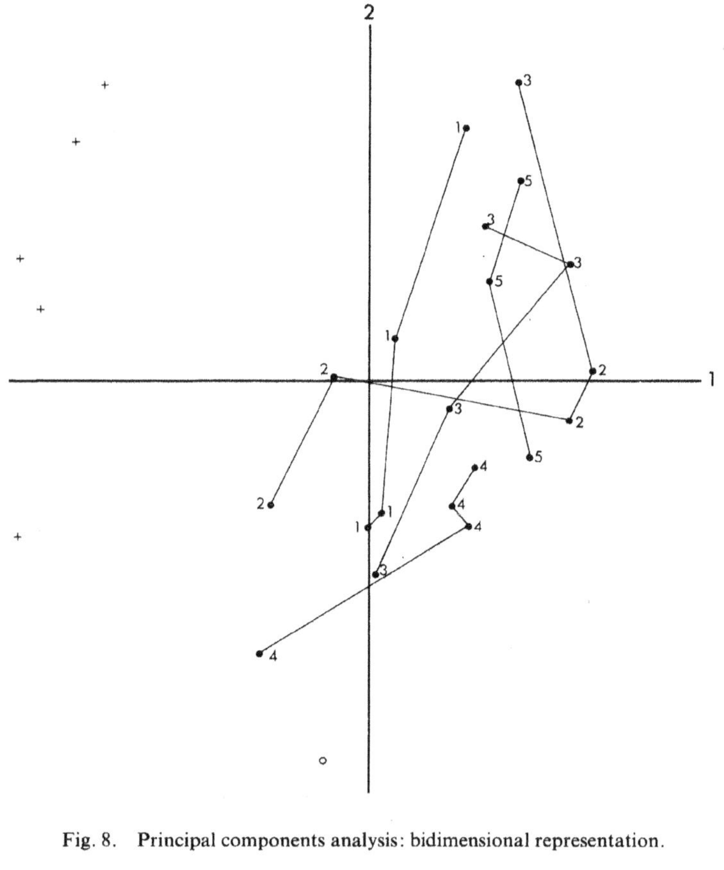

Fig. 8. Principal components analysis: bidimensional representation.

The first two principal components account for 58 % and 19 %, respectively, of total variance. The two morphotypes are well separated along the first axis (Figure 8). As might be expected from univariate analyses, most of the variation along this axis is attributable to cell size ($\overline{L}$) and peristome relative length ($\overline{P}$). No further structure of data is detectable. The 5 species of the *vannus* complex are not separated. As for clone

Frc 3, it appears somewhat marginal in the range of *vannus* morphotype, but it is not clearly dissociated from the complex.

3.6. Classification of clones by aggregative procedure

The same set of data was used for the construction of a dendrogram by an aggregative procedure, on the basis of euclidian distances in the space defined by the 6 standardized variables. The procedure includes a succession of steps. Each one consists in aggregating into a single group the closest two clones or groups of clones. For computation of distances a group is characterized by the gravity center of the representative points of the set of clones included in the group.

The first dichotomy of the resulting dendrogram (Figure 9) separates the two morphotypes. No clustering appears within any of them. The clone *Frc 3* is located close to clones which undoubtedly belong to the *E. vannus* complex.

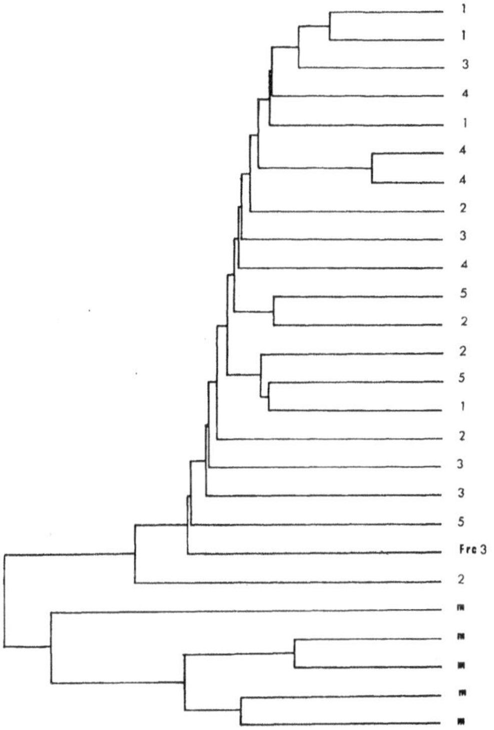

Fig. 9. Morphological relationships between 26 clones, visualized in the form of a dendrogram. The numbers refer to the species within *E. vannus* complex, the letter *m* refers to the *minuta* morphotype.

3.7. Attempt of species discrimination according to multivariate criteria

The above methods are not well suited to discrimination of very similar species. Thus a discriminant analysis was carried out. The principle is to determine new variables ('canonical variates'), which are linear combinations of the primary ones, in such a way that the ratio of interspecific to intraspecific variances is maximum. In the present study, a stepwise procedure was used, i.e. the primary variables are introduced in successive steps, following decreasing contribution to discrimination.

The analysis was restricted to the 20 clones of the above sample which undoubtedly belong to the *E. vannus* complex. The primary variables were introduced in the following order: $\overline{Q}, \overline{C}, \overline{P}, \overline{L}$, and $\overline{R}$. The first three canonical variates, accounting for 65, 28 and 4% of total variance, respectively, are:

$$X_1 = 16.608 - 0.171\,\overline{L} - 1.002\,\overline{R} + 6.935\,\overline{Q} - 0.515\,\overline{C} - 22.214\,\overline{P}$$

$$X_2 = 65.093 - 0.223\,\overline{L} - 0.520\,\overline{R} - 0.619\,\overline{Q} - 0.997\,\overline{C} - 70.608\,\overline{P}$$

$$X_3 = 25.472 - 0.0133\,\overline{L} - 0.980\,\overline{R} - 1.991\,\overline{Q} - 0.216\,\overline{C} - 3.208\,\overline{P}$$

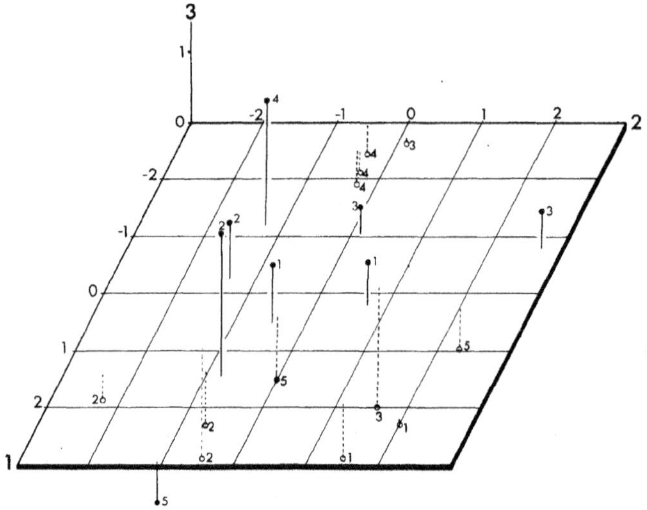

Fig. 10. Discriminant analysis within *E. vannus* complex: tridimensional representation. The numbers refer to the species.

A 3-dimensional representation of the results is given in Figure 10. Figure 11 gives a 2-dimensional representation in which both the *minuta* morphotype and *Frc 3* clones are represented. Concerning *E. vannus* complex, four distinct areas are shown in the plane X_1, X_2. One of them is common to species 1 and 5, while the three others are monospecific. The ranges of species 1 and 5 are separable when the variate X_3 is considered. Nevertheless, it cannot be concluded, in the present state of knowledge, that species discrimination can actually be achieved according to this procedure, since

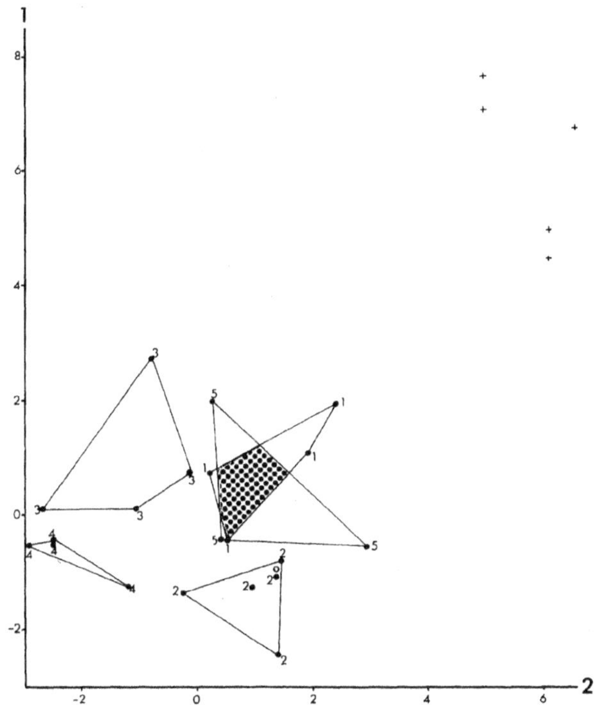

Fig. 11. Discriminant analysis: bidimensional representation. The *minuta* clones and *Frc 3* are represented in the referential defined for *E. vannus* complex (same symbols as in Figure 3).

the ranges are almost contiguous and the size of the sample is rather small.

In addition, it is clear that the variables X_1 and X_2 allow a fairly good discrimination of the two morphotypes. It can be seen in Figure 11 that the clone *Frc 3* is represented in the area of species 2; this holds true in 3-dimensional representation.

4. Discussion and Conclusions

4.1. Taxonomy and Nomenclature

It is clear from our data that the section *A* of genus *Euplotes* includes at least two morphotypes, readily discriminated on the basis of bidimensional analysis. Using our standardized culture and staining methods, when $\overline{L}$ (μm) is the mean cell length of a clone and $\overline{P}$ the mean relative peristome length, then the two morphotypes may be defined as follows (Figure 3):

$$\text{vannus}: \overline{P} + 0.006\,\overline{L} > 1.21$$
$$\text{minuta}: \overline{P} + 0.006\,\overline{L} < 1.11$$

They do not differ strikingly in other respects. The number of dorsal ciliary rows has no discriminative value, though the ranges of variation within the two morphotypes are somewhat different. Similar conclusions apply to the number of cilia per row and to the number of caudal cirri.

Each morphotype includes some 'conventional taxa': *vannus*, *crassus* and *mutabilis* for the former, *minuta* and *cristatus* for the latter. We retained for the designation of the morphotypes the earliest in each set of 'species' names.

No clone was found with the morphological attributes of *E. balticus*. According to Dragesco (1966), the most outstanding feature of this form is a very low number of dorsal rows: 6, a value which falls out of the range of our observations. Moreover, it may be inferred from the only drawing given by Dragesco that the number of cilia per row is rather low, too: about 12, while Curds (1975) gives in his key a much lower value: 7 to 10. Since *E. balticus* is a rather large form (60 to 100 μm: in the range of *vannus* morphotype, provided that a comparison with our metric data remains valid despite differences in culture methods), its dorsal ciliature appears as quite scarce. In the present state of our investigations, we cannot exclude that this form would be representative of a third morphotype.

The *vannus* morphotype includes at least 5 biological species. They are true sibling species or dualspecies (see Bernardi, 1980), since they are so closely similar at morphological level (and even at biochemical level: Machelon and Demar-Gervais, 1983) that the most powerful identification method remains the analysis of mating reactions with standard clones. According to our discriminant analysis, it may be expected that multidimensional techniques of diagnosis will be available in the future; however, the results presented here must be confirmed on the basis of more extensive sampling.

A few clones have been found which exhibited clearly the attributes of the *vannus* morphotype, but no mating reaction (or very weak reactions) with any of the standard clones. Such negative results do not give conclusive evidence for the discovery of a new species of the complex, since they may reflect one or another of a number of quite different situations, among which senility or sterility (in our breeding analyses, clones are occasionally obtained which, despite normal vegetative growth, never reach sexual maturity: such sterile clones are expected to occur in nature at low frequency). A new species can be described only when at least two clones prove to be sexually compatible with each other and incompatible with all representative clones of previously known species. One clone of this kind is described in the present paper, *Frc 3*, the taxonomic status of which remains obscure. Since it seems related to the *minuta* morphotype according to electrophoretical patterns, we have logically excluded it from the *vannus* complex. This observation raises some questions concerning the evolutionary mechanisms involved in diversification within *Euplotes*, section *A*. This point will be discussed later.

The morphotype *vannus*, including clones, such as *Frc 3*, of uncertain specific status, fits exactly to the concept of *supraspecies* recently introduced by Génermont and Lamotte (1980). This taxonomic unit is useful since it is the only one which can be considered by most zoologists or unspecialized protozoologists who must not be dependent on sophisticated identification procedures (such as confrontations of clones with standard stocks). It is nevertheless important to remember that the supraspecies cannot be integrated into a hierarchical taxonomic system, because it is not necessarily monophyletic (while the strictly speaking *E. vannus* complex is presumably monophyletic).

Thus, in our present state of knowledge, it can be stated that the section *A* of *Euplotes* includes two supraspecies: *E.* suprasp. *vannus*, including 5 sibling species together with some other forms of uncertain status, and *E.* suprasp. *minuta*, the structure of which remains unknown. The notations used here are in agreement with the suggestion of Bernardi (1980) for superspecies and ultraspecies. As for the taxonomic unit which has been referred to as *E. vannus* complex, another possible notation is *E.* ultrasp. *vannus*.

A further task should be to give species names to each of the 5 sibling species of the *E. vannus* complex. Such a task has been achieved by Sonneborn (1975) for the *Paramecium aurelia* complex and by Nanney and MacCoy (1976) for the *Tetrahymena pyriformis* one. In the present case, some further data still seem necessary, especially in the field of diagnostic procedures.

4.2. Evolutionary mechanisms

The 5 species recognized up to now within the *E. vannus* complex are closely similar not only at morphological level but also at biological and biochemical level (Machelon, 1982, 1983; Machelon and Demar-Gervais, 1983). It is thus likely that these species originated recently from their last common ancestor. From this point of view, the *E. vannus* complex seems quite different from '*Tetrahymena pyriformis*' in which the genetic interspecific distances, estimated from various biochemical data, are very large, so that it is assumed that the divergence between species has been carried out a long time, but has been restricted by some morphological constraints (Nanney, 1980). The status of '*Paramecium aurelia*' seems to be intermediate.

Thus it appears that the various complexes of sibling species which are presently known among Ciliates illustrate a wide range of situations, showing clearly that there is no strict correlation between the degree of morphological differentiation and the length of the divergence period.

It is obvious that speciation may occur without any perceptible morphological change. This fact is now well known and its direct consequence is the common occurrence of sibling species, not only among protists, but also in many groups of animals and plants. Reciprocally, when some morphological diversification is observed within a monophyletic group, it must be questioned if this diversification is associated or not with speciation processes. The Ciliates provide some information on this point, when closely related complexes of sibling species are compared. For instance, both '*Paramecium aurelia*' and '*P. caudatum*' are complexes of numerous species (15 are known in the former, with probably even more in the latter). In each of these complexes the component species exhibit a very high degree of morphological similarity, but there is a very strong discontinuity between the two complexes. It seems rather surprising that, in spite of a rather high rate of speciation the result of which is the high numbers of sibling species, no species exhibit intermediate characters between the two morphotypes. This observation strongly suggests, in our minds, that two distinct evolutionary processes are involved in diversification within *Paramecium*: (*i*) speciation without morphological change and (*ii*) jump from one morphotype to another; the latter could be a rather rapid process: the observed morphotypes would represent stable (at evolutionary level) morphological states, each one associated with a definite adaptive zone, while intermediate morphological states would be unstable.

The situation in section *A* of *Euplotes* seems similar. A clear discontinuity (though less conspicuous than in *Paramecium*) does occur between *vannus* and *minuta* morphotypes. In this case as well, two kinds of evolutionary processes are likely. Speciation without morphological change is examplified by the 5 sibling species of the *E. vannus* complex. Morphological changes occurred at least twice: in a first step, the two morphotypes must have emerged from their common ancestor (we cannot know whether this process has been gradual or saltatory); later, a jump from *minuta* to *vannus* would have generated a form morphologically similar to *vannus* but phylogenetically closer to *minuta*, a representative of which would be the *Frc 3* clone. Since a jump from one morphotype to another is probably associated with a jump from one adaptive zone to another, it is likely that it implies the development of some reproductive barrier and thus is associated with speciation.

This hypothetical mechanism calls for some comments.

1. The jump from one morphotype to another among a limited array of stable states can eventually result in apparent exceptions to the dogma of irreversibility of evolutionary processes; in fact it is not a true reversion, since all selectively neutral characters (for instance at least a part of electrophoretical patterns) are not reversed.
2. The postulated mechanism is reminiscent of the 'punctuated equilibria' model developed by Eldredge and Gould (1972), but these authors use the word speciation for a process which implies some morphological change.
3. Further data are necessary in order to test our hypothesis. Among other tasks, the genetic and taxonomic structure of the *minuta* morphotype must be ascertained and discrepancies between morphological, biological and biochemical classifications must be carefully analysed.

References

Bernardi, G.: 1980, 'Les Catégories Taxonomiques de la Systêmatique Évolutive', in C. Bocquet, J. Génermont and M. Lamotte (eds.), *Les Problêmes de l'Espèce dans le Règne Animal*, vol. 3, *Mém. Soc. Zool. Fr. no 40*, 373–425.

Chatton, E. and Lwoff, A.: 1930, 'Imprégnation, par Diffusion Argentique, de l'Infraciliature des Ciliés Marins et d'Eau Douce, Après Fixation Cytologique et sans Desiccation', *C. R. Soc. Biol.* **104**, 834–836.

Curds, C. R.: 1975, 'A Guide to the Species of Genus *Euplotes* (Hypotrichida, Ciliatea)', *Bull. Brit. Mus. (Nat. Hist.). Zool.* **28**, 1–61.

Dragesco, J.: 1966, 'Observations sur Quelques Ciliés Libres', *Arch. Protistenk.* **109**, 155–206.

Eldredge, N. and Gould, S. J.: 1972, 'Punctuated Equilibria: An Alternative to Phyletic Gradualism', in T. J. M. Schopf (ed.), *Models in Palaeobiology*, pp. 85–115.

Frankel, J.: 1973, 'Dimensions of Control of Cortical Patterns in *Euplotes*: The Role of Preexisting Structures, the Clonal Life Cycle and the Genotype', *J. Exp. Zool.* **183**, 71–94.

Frankel, J.: 1975, 'An Analysis of the Spatial Distribution of Ciliary Units in a Ciliate, *Euplotes minuta*', *J. Embryol. Exp. Morphol.* **33**, 553–580.

Frankel, J. and Heckmann, K.: 1968, 'A Simplified Chatton-Lwoff Silver Impregnation Procedure for Use in Experimental Studies with Ciliates', *Trans. Amer. Microsc. Soc.* **87**, 317–321.

Gates, M. A.: 1977, 'Application of the Analysis of Positional Information to the Cirral Patterns of the Ciliate Genus *Euplotes*', *Nature* **268**, 362–364.

Gates, M. A. and Berger, J.: 1976, 'Morphological Inseparability of *Paramecium primaurelia* and *Paramecium pentaurelia*', *Trans. Amer. Microsc. Soc.* **95**, 507–514.

Gates, M. A., Powelson, E. E., and Berger, J.: 1974, 'Syngenic Ascertainment in *Paramecium aurelia*', *Syst. Zool.* **23**, 482–489.

Génermont, J.: 1976, 'Le Problême de l'Espèce chez les Protozoaires' in C. Bocquet, J. Génermont, and M. Lamotte (eds.), *Les Problêmes de l'Espèce dans le Règne Animal*, vol. 1, *Mém. Soc. Zool. Fr. no 38*, 375–407.

Génermont, J. and Lamotte, M.: 1980, 'Le Concept Biologique de l'Espèce dans la Biologie Contemporaine' in C. Bocquet, J. Génermont, and M. Lamotte (eds.), *Les Problêmes de l'Espèce dans le Règne Animal*, vol. 3, *Mém. Soc. Zool. Fr. no 40*, 427–452.

Génermont, J., Machelon, V., and Tuffrau, M.: 1976, 'Données Expérimentales Relatives au Problême de l'Espèce dans le Genre *Euplotes* (Ciliés Hypotriches)', *Protistologica* **12**, 239–248.

Hallet, M. M.: 1978, 'Recherches Préliminaires sur la Composition Spécifique d'un Peuplement d'*Euplotes* du Groupe *vannus*', *J. Protozool.* **25**, 55 A.

Laloe, F.: 1979, 'Contribution à l'Étude de la Variabilité Intraspécifique pour le Nombre de Cinéties Dorsales et de Cirres Caudaux chez une Espèce du Complexe *Euplotes vannus* (Ciliés Hypotriches)', *Arch. Zool. Exp. Gén.* **120**, 109–120.

Machelon, V.: 1978, 'Étude d'une Troisième Espèce dans le Complexe *Euplotes vannus* (Ciliés Hypotriches)', *Protistologica* **14**, 15–22.

Machelon, V.: 1982, 'Données sur le Complexe *Euplotes vannus*: l'Existence d'une Cinquième Espèce, Analyse des Mécanismes d'Isolement Reproductif', *Protistologica* **18**, 345–354.

Machelon, V.: 1983, 'Étude du Complexe *Euplotes vannus* (Ciliés Hypotriches)', Thèse d'Etat, Université de Paris-Sud.

Machelon, V. and Demar-Gervais C.: 1983, 'Electrophoretic Variation among Genus *Euplotes* (Ciliata, Hypotrichida): Comparative Data for the Sibling Species Complex *Euplotes vannus* and Survey of Intrageneric Variability', *J. Protozool.* **30**, in press.

Nanney, D. L.: 1980, *Experimental ciliatology*, J. Wiley and Sons, XII + 304 p.

Nanney, D. L. and MacCoy, J. W.: 1976, 'Characterization of Species of the *Tetrahymena pyriformis* Complex', *Trans. Amer. Microsc. Soc.* **95**, 664–682.

Nyberg, D.: 1981, 'Three New "Biological" Species of *Tetrahymena* (*T. hegewishi* n. sp., *T. sonneborni* n. sp., *T. nipsingi* n. sp.) and Temperature Tolerance of Members of the "*pyriformis*" complex', *J. Protozool.* **28**, 65–69.

Sonneborn, T. M.: 1975, 'The *Paramecium aurelia* Complex of 14 Sibling Species', *Trans. Amer. Microsc. Soc.* **94**, 155–178.

Tuffrau, M.: 1960, 'Révision du Genre *Euplotes*, Fondée sur la Comparaison des Structures Superficielles', *Hydrobiologica* **15**, 1–77.

AN EVALUATION OF A POSSIBLE PHYLOGENETIC RELATIONSHIP BETWEEN THE EUGLENOPHYTA AND KINETOPLASTIDA

PETER A. KIVIC and PATRICIA L. WALNE

Department of Botany, University of Tennessee, Knoxville, TN 37996-1100, U.S.A.

(Received 5 October, 1983)

Abstract. Without close phylogenetic ties to any of the other algae, the Euglenophyta are a taxonomic enigma. The argument is made here that the euglenophytes have extensive morphological homology with the zooflagellate trypanosomatids, bodonids, and *Isonema*. A phylogenetic sequence is postulated linking these groups; the available data suggest that the euglenophytes had their origin in these zooflagellates and that the introduction of the chloroplast and a phototrophic mode of nutrition was a relatively late step in this evolutionary process.

1. Introduction

The euglenoid flagellates are a traditional taxonomic puzzle: a diverse collection of phototrophs, phagotrophs and osmotrophs which are all closely related to one another but not, according to all available evidence, to any other algal group (Dodge, 1975; Leedale, 1967). Their closest affinities are perhaps with some fairly obscure zooflagellates. The present work is an exploration of these possible euglenoid-zooflagellate connections.

Three zooflagellate taxa were examined for possible phylogenetically relevant similarities with the euglenoids: the trypanosomatids, uniflagellate and exclusively parasitic; the bodonids, tiny, biflagellate and mostly free living; and the biflagellate marine flagellate, *Isonema*. The bodonids and trypanosomatids are closely related and have been combined as the Kinetoplastida (Brooker, 1971a; Honigberg, 1963); together or separately, they have been suggested a number of times as possible euglenoid relatives (e.g., Bovee, 1972; Brooker, 1971a; Burzell, 1973; Dodge, 1973; Hutner *et al.*, 1968; Leedale, 1970; Mignot, 1964, 1966; Porter, 1973; Round, 1980; Schuster *et al.*, 1968; Swale, 1973; Taylor, 1976, 1980; Vickerman and Preston, 1976). *Isonema* has no known affinities, but has been described as having certain euglenoid properties and possible euglenoid affiliations (Porter, 1973; Schuster *et al.*, 1968; and a presumptive *Isonema*: Bodammer and Sawyer, 1981).

Information obtained from an extensive literature search has afforded a critical comparison of several organellar systems in these organisms.

2. Morphological Comparisons

The above groups are by no means identical, and within each there are, of course,

Origins of Life **13** (1984) 269-288. 0302-1688/84/0133-0269$03.00.

instances of broad morphological variation among taxa; but we do find a relatively large number of characters which are similar or essentially the same in most of the organisms considered:

a. *General morphology*. The kinetoplastids, euglenoids and *Isonema* all manifest a similar external morphology: an ovoid-to-lanceolate body shape, usually with a helical symmetry; the flagella (usually two) insert within an anterior subapical pocket. In the phagotrophs, a tubular ingestion apparatus usually opens near the right-ventral edge of the flagellar pocket.

b. *Pattern of movement*. A peculiar crawling movement termed 'metaboly' or 'euglenoid motion' is found in all euglenoids not totally restricted by a rigid pellicle, and has traditionally been considered a basic euglenoid characteristic (Leedale, 1967; Mikolajczyk and Kuznicki, 1981; Pringsheim, 1956). Euglenoid motion, or behavior very similar to it, however, has also been described in *Isonema* (Porter, 1973; Schuster *et al.*, 1968), and in the bodonids *Cryptobia vaginalis* (Vickerman, 1977) and *Rynchomonas metabolita* (Burzell, 1973).

c. *Flagellar placement*. Of the two flagella, usually only the anterior is employed in propelling the cell. The posterior flagellum is very reduced in some phototrophic and osmotrophic euglenoids (Leedale, 1967; Leedale and Hibberd, 1974), but in the phagotrophs it is characteristically long and trails passively as a ventral 'skid' in contact with the substratum (Burzell, 1973; Leedale, 1967; Mignot, 1966, 1975; Swale, 1973; Vickerman, 1977; Vickerman and Preston, 1976), possibly serving to keep the cell close to its food source (Burzell, 1973). The single motile flagellum of the parasitic trypanosomatids is probably the posterior one (Kivic and Walne, unpublished).

d. *Paraflagellar rod*. In addition to the usual 9 + 2 microtubular axoneme the euglenoid flagellum contains a paraflagellar rod (PFR, paraxial rod, intraflagellar structure) in contact with the axoneme and with a distinctive latticework substructure (Hyams, 1982; Kivic and Vesk, 1972; Leedale, 1967; Mignot, 1966; Moestrup, 1982). A similar PFR structure is found in most trypanosomatids (Fuge, 1969; McGhee and Cosgrove, 1980) and in all of the bodonids thus far examined ultrastructurally. The PFR is absent from *Isonema* (Porter, 1973; Schuster *et al.*, 1968), a few of the trypanosomatid species (Freymuller and Camargo 1981; McGhee and Cosgrove, 1980), and the reduced posterior flagella of numerous euglenoids (Leedale, 1967).

e. *Nucleus and nuclear division*. Nuclear architecture and division in euglenoids (Leedale, 1970) are essentially the same as in bodonids, trypanosomatids and *Isonema* (Raikov, 1978), which also have a totally closed mitosis, an intranuclear spindle, and no metaphase plate formation. In all these taxa, cytokinesis begins at the flagellar pocket and proceeds posteriorly, following the helical organization of the cell cortex (Grasse, 1952; Hollande, 1952a, b; Leedale, 1967; Porter, 1973; Schuster *et al.*, 1968).

f. *Mitochondria*. The euglenoid chondriome (at least as represented by that of *Euglena gracilis*) also appears to be a single large reticulum (Pellegrini, 1980). In both the bodonids and the trypanosomatids, the chondriome similarly consists of a single large mitochondrion (Paulin, 1975, 1977), which varies in shape from a simple loop (*Rynchomonas*, *Bodo*) or tube (*Trypanosoma*) to more complex ramifying designs

(*Cryptobia*) (Brugerolle *et al.*, 1979; Vickerman, 1977). One common feature of their giant mitochondria serves as the distinctive characteristic linking the bodonids and trypanosomatids as the group Kinetoplastida: a prominent mDNA mass, the kinetoplast, usually occurs immediately adjacent to the flagellar base (Brooker, 1971a; Honigberg, 1963).

The euglenoids and kinetoplastids are included in that group of protists with flat, rather than tubular, cristae (Leedale, 1967; Taylor, 1974), a characteristic reportedly of phylogenetic significance (Stewart and Mattox, 1980). The euglenoid cristae are constricted at their bases, a character taken to be diagnostic of this group (Leedale, 1967); however, constricted bases appear to be typical kinetoplastid cristae as well (Taylor, 1976). This particular characteristic may not be apparent, however, unless the mitochondrion is in a relatively active metabolic state. For example, the cristae become sparse and tubular in *Trypanosoma brucei* when in the bloodstream phase, in which the Krebs cycle is inactive (Vickerman, 1971), in *Phytomonas davidi* in its latex phase (Postell and McGhee, 1981), and in *Bodo caudatus* entering encystment (Brooker and Ogden, 1972). This may explain why the small, 'vestigial-looking' (Porter, 1973) mitochondria of *Isonema* have sparse, tubular cristae (Porter, 1973; Schuster *et al.*, 1968), or why the *Isonema-like* organism described by Bodammer and Sawyer (1981) has shelf-like cristae.

g. *Contractile vacuole*. A 'Type D' (after Patterson, 1980) contractile vacuole is found in euglenoids, most bodonids, and even most trypanosomatids, albeit they are endoparasites (e.g., Brooker, 1971b; Clark, 1959; Vickerman, 1963). *Isonema*, like most marine flagellates, has no contractile vacuole (Porter, 1973; Schuster *et al.*, 1968). In each case, the contractile vacuole empties into the bottom of the flagellar pocket and is formed anew by the fusion of smaller accessory vacuoles; the source of the latter appears to be a nearby Golgi apparatus.

h. *Trichocysts*. These are not universal but are found only in *Isonema* and some euglenoids. *Peranema* and *Entosiphon* contain batteries of trichocysts, peripheral ejectile organelles which, upon irritation, discharge violently (Mignot, 1966; Mignot and Hovasse, 1973; Hilenski and Walne, 1983). They are hollow cylinders with a distinctive basketweave substructure (Hilenski and Walne, 1983; Mignot, 1966); they are obviously different from the various other types of ejectile organelles (e.g., ejectosomes, discobolocysts) found in other algae (Dodge, 1973). *Isonema nigricans* has batteries of ejectile organelles also; these are reportedly similar upon discharge to the *Entosiphon* trichocysts (Schuster *et al.*, 1968). In their undischarged state the *Isonema* trichocysts are identical to neither those of *Peranema* nor *Entosiphon*, but they resemble each of them far more closely than do those of any other protists (e.g., Figure 1). Neither *I. papillatum* (Porter, 1973) nor any of the kinetoplastids have been reported to have ejectile organelles of any kind.

i. *Flagellar roots*. The bodonids have three flagellar roots, which extend up the sides of the flagellar pocket and, as will be discussed below, are also associated with other structural elements (Figures 2 and 3; and Brugerolle *et al.*, 1979). The euglenoids also have three microtubular bands associated with the reservoir (e.g., Mignot, 1966;

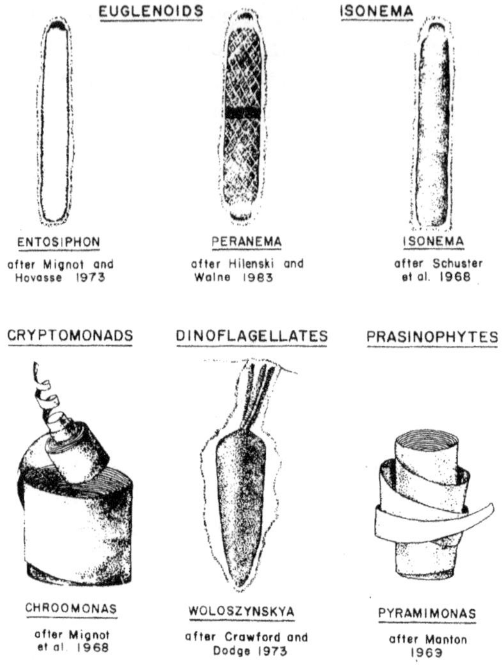

Fig. 1. Comparison of undischarged trichocysts of *Isonema*, euglenoids and other algae. The *Isonema* trichocysts resemble those of the euglenoids but not those of cryptomonads, dinoflagellates and prasinophytes.

Schuster and Hershenov, 1974); in *Menoidium*, *Rhabdomonas* (Leedale and Hibberd, 1974), *Euglena* (Piccini and Mammi, 1978) and *Trachelomonas* (West *et al.*, 1980), these are seen to originate at the basal bodies. *Isonema* has three flagellar pocket bands, too, at least one of which originates at a basal body (Porter, 1973). Only the uniflagellate trypanosomatids (Postell and McGhee, 1981) and the uniflagellate euglenoid *Scytomonas* (Mignot, 1966) differ from this pattern, having only one flagellar root.

Altogether, Brugerolle *et al.* (1979) identify five basic microtubular components in the bodonid cytoskeleton: the three flagellar roots and two others, as detailed in Figure 2. Two of these are particularly relevant to the present discussion because they can also be tentatively identified in some or all of the other taxa under consideration here.

The dorsal fibril (*Fd*) is the major component of the bodonid peripheral cytoskeleton, while the reinforced-microtubule root (*mtr*) also forms part of the structural support of the ingestion apparatus. Fortuitously, both of these microtubular bands are readily identifiable in electron micrographs because each carries a characteristic 'signature': the *Fd* microtubules are all linked together to form a continuous sheet, whereas the *mtr* microtubules are individually tied to the overlying cell membrane by fibrous struts (Figure 2).

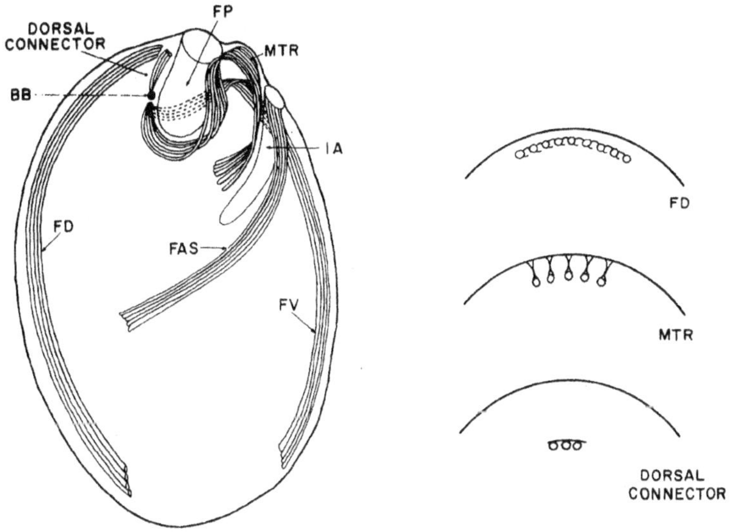

Fig. 2. Microtubular bands of bodonids (after Brugerolle *et al.*, 1979). Left: Of the five microtubular bands which have been described, two are of particular relevance to the present discussion: The *Fd* originates near the anterior end of the organism and proceeds posteriorly along the dorsal surface; the *mtr* originates at the basal body of the posterior flagellum and follows a convoluted pathway (cf. Figure 3). IA = ingestion apparatus; FP = flagellar pocket; BB = basal body (kinetosome). Right: certain of these microtubular bands typically bear accessory structures which can serve as identifying characteristics: the *Fd* microtubules carry two side arms; the dorsal side arms link the microtubules together to form a continuous sheet; the *mtr* microtubules are individually bound to the cell membrane by fibrous connectives; the dorsal connector microtubules are associated with an overlying fibrous lamella*.

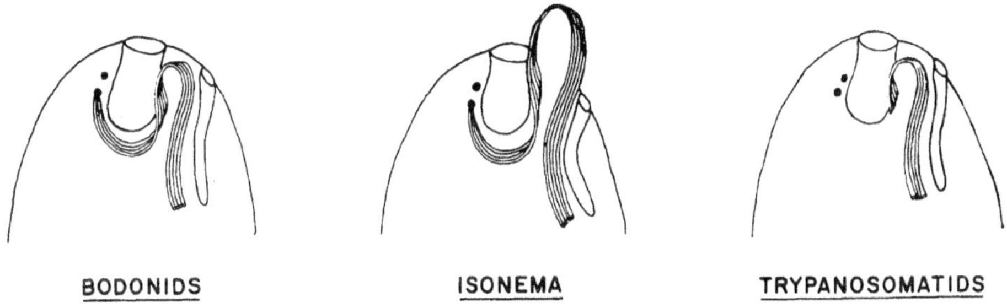

Fig. 3. The *mtr*: This microtubular band is both a flagellar root and a structural component of the ingestion apparatus. Originating at the basal body of the posterior flagellum, the *mtr* curves around the bottom of the flagellar pocket, makes a U-turn inward at the cell surface to form part of the ingestion apparatus structural support. In some species, the especially prominent U-turn forms a preoral crest or probscis; it may be partially covered with a patch of glycoprotein filaments ('hairy tomentum'). The *mtr* can be identified in bodonids, trypanosomatids and *Isonema* but not, as yet, in euglenoids.

* The figures are diagrammatic representations of particular structures in the several groups that we are comparing; they are not intended to be exact, detailed depictions of any particular organisms.

3. Structural Homologies

We also find three other structural characteristics which are different, but recognizably homologous, in the various taxa. Since these structures are of particular importance to our argument, they will be described here in greater detail:

a. *The peripheral cytoskeleton and associated structures*. Some of the bodonids (e.g., *Rynchomonas*: Swale, 1973; *Bodo*: Brooker, 1971a; Pitelka, 1963b; *Cephalothamnium*: Hitchen, 1974) have a very simple cytoskeleton which consists of isolated microtubular bands that support mainly the anterior part of the cell (Brugerolle *et al*., 1979); the major peripheral element is the dorsal fibril (*Fd*), with perhaps a minor contribution from the ventral fibril (*Fv*) (Figure 4). The bodonid Fd also has an identifying 'signature': the microtubules are cross-linked and form one broad microtubular sheet (Figure 4) (Brugerolle *et al*., 1979). Other bodonids (e.g., *Cryptobia*: Brugerolle *et al*., 1979; *Icthybodo*: Joyon and Lom, 1966, 1969; *Spiromonas*: MacDonald and Darbyshire, 1977; or perhaps *Dimastigella*: see Brugerolle *et al*., 1979) have a more complex cytoskeleton consisting of an almost-complete helical 'corset' of linked anterio-ventral microtubules surrounding the cell; this 'corset' appears to be a greatly expanded *Fd* (Brugerolle *et al*., 1979).

The same interpretation can be applied to the peripheral cytoskeletons of the trypanosomatids, *Isonema*, and perhaps the euglenoids. In each of these, the cytoskeleton consists entirely or in part of a helical peripheral sheet, or discontinuous bands, of linked anterio-ventral microtubules.

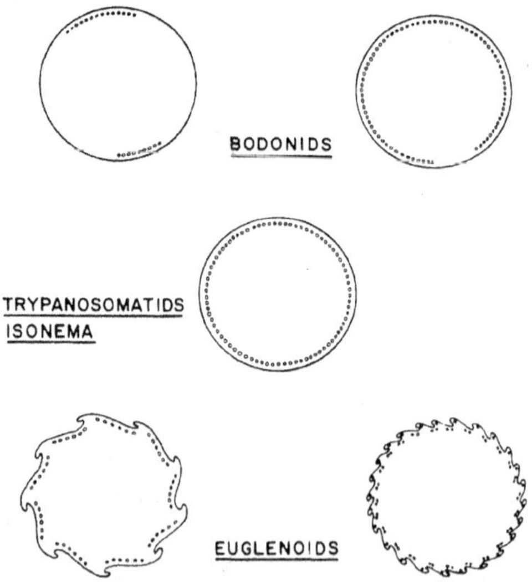

Fig. 4. Peripheral cytoskeletons of bodonids, trypanosomatids, *Isonema* and euglenoids in diagrammatic transverse section. The simplest bodonid cytoskeleton is composed of a large *Fd* band and a smaller *Fv* band. The more complex, nearly complete microtubular 'corset' of the other bodonids, and the complete 'corsets' of trypanosomatids and *Isonema*, probably include both of these components. The similar-appearing euglenoid cytoskeleton is associated with an overlying pellicle.

Isonema and the trypanosomatids have helical microtubule corsets (Angelopoulus, 1970; Porter, 1973; Schuster *et al.*, 1968), in which the microtubules are cross-linked (*Isonema*: see Fig. 22 in Porter, 1973; trypanosomatids: DeSouza *et al.*, 1976; Linder and Staehlin, 1977; Paulin and McGhee, 1971; Postell and McGhee, 1981); these can be most simply described as more complete versions of the *Cryptobia*-type cytoskeleton (Figure 4).

The euglenoids also have a peripheral helical layer of microtubules; in this instance however, they are divided into a series of bands around the periphery, with each band overlain by a strip of the unique euglenoid pellicle (Figure 4). The numbers of pellicle strips and microtubules vary widely among euglenoids. Most phagotrophs (Heteronematales, Sphenomonadales) have few pellicle strips, but many microtubules associated with each strip; the peripheral microtubular layer can appear almost continuous around the cell perimeter and shows a distinct resemblance to the microtubular layer in *Isonema*, trypanosomatids and some bodonids (Figure 4). Phototrophs (Euglenales, Eutreptiales) have many pellicle strips but only a small number of microtubules associated with each strip (Figure 4). The (non-phagotrophic, non-phototrophic) Rhabdomonadales have a small number of fused strips, each associated with a small number of microtubules. The pellicle strips, and some of their associated microtubules, proceed over the lip of the flagellar pocket and down the narrow distal canal; in this region, at least, the microtubules are clearly seen to be linked as bands of two or three microtubules (Leedale and Hibberd, 1974; West *et al.*, 1980) or as a continuous linked sheet (Dodge, 1975; Piccini and Mammi, 1978). The position, orientation, and cross-linking of these microtubules, therefore, all contribute to the conclusion that they are homologous with the microtubular corset and *Fd* band of the bodonids.

b. *The structural support of the ingestion apparatus. Isonema*, most bodonids, and some euglenoids are predators. In every such species investigated thus far, there is a permanent ingestion apparatus; it is always in the same location and of similar appearance. It is typically a tubular invagination near the right-ventral lip of the flagellar pocket, supported by a number of parallel microtubular structures. In some bodonids (*Rynchomonas*, *Spiromonas*, *Icthyobodo*, *Bodo caudatus*: Burzell, 1975; Joyon and Lom, 1966; MacDonald and Darbyshire, 1977) the ingestion apparatus projects forward as a rostrum or proboscis, but the anatomical relationships remain the same.

Many trypanosomatids also have an ingestion apparatus: in some species the opening is displaced to within the flagellar pocket, but the structure remains functional (Brown *et al.*, 1965; Gardner, 1974; Meyer and DeSousa, 1973; Preston, 1969; Steinert and Novikoff, 1960).

The structural support of the ingestion apparatus ranges from very simple in e.g., *Rynchomonas* (Burzell, 1973; Swale, 1973) and the trypanosomatids (e.g., Brooker, 1971b; Lom *et al.*, 1980; Preston, 1969; Steinert and Novikoff, 1960), to extremely complex in the euglenoid phagotrophs (Figure 5). In all cases, however, it is found to

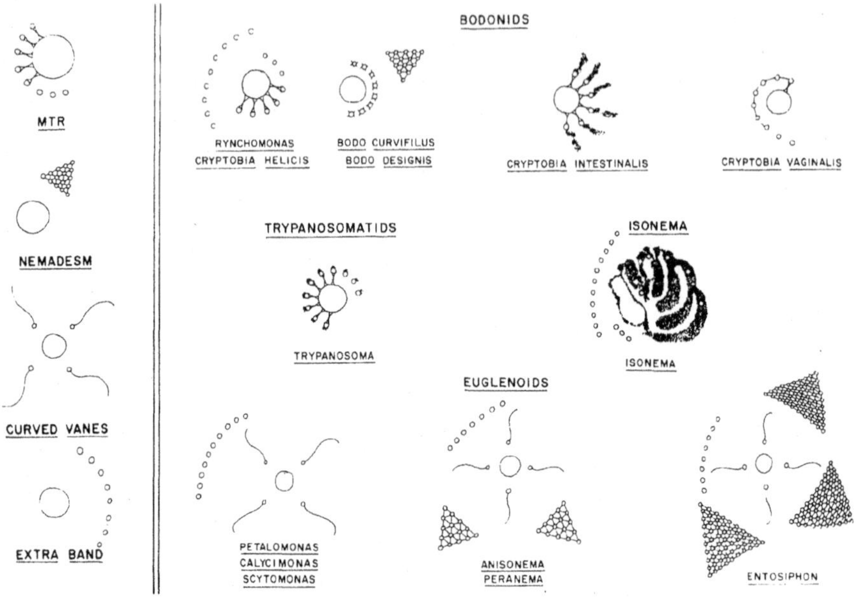

Fig. 5. Ingestion apparatus in transverse section. Left: The ingestion apparatus is composed of four basic microtubular components. Right: These basic components are combined and modified in a variety of ways in the many phagotrophs discussed, but the basic structural similarities are still apparent.

consist of one or more of four basic elements, in various combinations (Figure 5). These are:

(1) *The mtr*. Previously identified as one of the three bodonid flagellar roots, the *mtr* and its associated fibrous connectives form the chief structural component of the ingestion apparatus in the bodonids (Brugerolle *et al*., 1979); it can be identified in most bodonids for which there are electron micrographs by its position and/or the distinctive fibrous struts binding its microtubules to the overlying plasmalemma. Originating as a flagellar root, the *mtr* runs under the flagellar pocket to the right-ventral lip, where it makes a prominent U-turn back into the cell and forms the main structural support of the ingestion apparatus; within the ingestion apparatus, these microtubules run parallel, and immediately adjacent to, the invagination. At the place where the *mtr* makes its U-turn, the cell membrane bulges out as a projecting 'preoral crest' (Brugerolle *et al*., 1979) (Figure 3). This structure is especially prominent, and coated with fine glycoprotein hairs (a 'hairy tomentum'), in *Cryptobia intestinalis* (Brugerolle *et al*., 1979), *C. vaginalis* (Vickerman, 1977), and *Bodo designis* (Eyden, 1977).

The simplest ingestion apparatus is perhaps that of the trypanosomatids, which consists simply of an invagination of the body or flagellar pocket wall surrounded by four to eight microtubules (depending on species) (Figure 5). In *Trypanosoma danilewskyi* it has now been established that at least some of these microtubules are homologous with the bodonid *mtr* band, even though they originate in this species, not as a flagellar root, but from a region just within the flagellar pocket (Lom *et al.*, 1980) (Figure 3).

In *Isonema* one of the three flagellar roots extends up the flagellar pocket wall, turns back over its ventral lip, and forms part of the ingestion apparatus cytoskeleton (Figure 3). Its eight or so microtubules are attached to the cell membrane by rows of fibrous connectives, some of which become quite enormous within the ingestion apparatus, forming the prominent 'semi-rosette' structure (Porter, 1973; Schuster *et al.*, 1968) (Figure 5).

The prominent anterior papilla of *I. papillatum* is a projection of the ventral flagellar pocket lip immediately anterior to the ingestion apparatus; it is formed by a large loop of this microtubular band (Porter, 1973). The papilla thus appears to be a very big preoral crest. The evidence for the homology of this microtubular band with the bodonid *mtr* is fairly convincing.

While the identification of an *mtr* homologue in euglenoids is not presently possible, a 'hairy tomentum' is associated with the ingestion apparatus orifice in *Anisonema* and *Entosiphon* (Mignot, 1966) and near the canal-reservoir junction in several green euglenoids (Dynesius and Walne, 1975; Rosowski and Lee, 1978; Willey *et al.*, 1973).

(2) *The nemadesm.* A stiff straight bundle, approximately triangular in cross section, of cross-linked microtubules, the nemadesm also runs parallel to the cytopharynx (Mignot, 1966) (Figure 5).

The prominent ingestion rods of the *Heteronematales*, (Christen, 1959; Leedale, 1967) are a pair of large nemadesms flanking the rest of the ingestion apparatus (Hilenski and Walne, 1980; Mignot, 1966; Nisbet, 1974). Three large nemadesms fused together at their larger anterior ends form the 'siphon' characteristic of *Entosiphon* (Leedale, 1967; Mignot, 1963, 1966). A pair of smaller nemadesms (30-50 microtubules each, versus 150-200 microtubules in *Entosiphon* and *Peranema*) was described in *Anisonema* (Mignot, 1966). The euglenoid nemadesms are evidently moveable, and their use as versatile cutting, tearing, and swallowing tools has been described (Chen, 1950; Leedale 1967; Nisbet, 1974).

Bodonids also have nemadesms, although they have not been described as such. They are much smaller than those of the euglenoids, and occur singly, rather than in pairs. The 'rod-organ' of *Bodo designis* (Eyden, 1977) and the 'microtubular prism' of *B. curvifilus* (Burzell, 1975) are straight bundles of 19 and 15, respectively, cross-linked microtubules; these bundles are approximately triangular in cross-section, and parallel the ingestion apparatus in each case. The smallest nemadesm is the 'quintubular fiber' of *Cephalothamnium* (Hitchen, 1974), consisting of only five cross-linked microtubules (two over three, in cross-section) running parallel to the cytopharynx. No function has yet been ascribed to the bodonid nemadesms.

(3) *Curved vanes.* A set of four (usually) long vanes radiate from the tubular cytopharynx; in cross-section they are curved, somewhat resembling the blades of a pinwheel, with each having a microtubule along the edge facing the lumen (Mignot, 1966) (Figure 5).

Whereas the bodonids, trypanosomatids, and *Isonema* have the *mtr* structure immediately surrounding the lumen of the ingestion apparatus, in the phagotrophic euglenoids this position is occupied by the vane assembly. This is seen in various degrees of development in *Scytomonas, Petalomonas, Calycimonas, Entosiphon*, and *Peranema* (Mignot, 1966); in the last two, the posterior parts of the vanes appear to be fused to the nemadesms (Mignot, 1966; Nisbet, 1974). While no homology between the vane assembly and the *mtr* has been established, it may be noted that both consist of long parallel microtubules with large fibrous appendages.

(4) *An extra band of microtubules.* In some cases the ingestion apparatus structural support includes one or more unidentifiable additional microtubular bands of unknown origin that run parallel to the rest of the structures (Figure 5). An especially large one is evident in *Isonema* (Porter, 1973; Schuster *et al.*, 1968), and smaller ones are seen in some euglenoids (Mignot, 1966).

c. *The flagellar transition zone.* Separating the emergent flagellum from the basal body is this short segment with a distinctive morphology. Transition zones among organisms vary considerably in structural detail (Dodge, 1973). It is of special interest, therefore, that the transition zone of all of the organisms under consideration here share a common pecularity: the central area is electron-transparent and devoid of discernible structure (Figure 6).

End-plates (which appear to be complete septa in some cases, but are open annuli in most) delimit the proximal and/or distal ends of the transition zone in trypanosomatids, most bodonids, and *Isonema*, but not euglenoids. The axonemal central microtubules terminate near the proximal end of the transition zone; the nature of this terminus varies.

In trypanosomes, one (Anderson and Ellis, 1965) or perhaps both (Brack, 1968) of the central tubules extend through the distal plate and terminate just proximal to it. In those bodonids in which the transition zone has end-plates, the central tubules appear to terminate at a small toroidal structure (axosome) just distal to the distant end-plate. A small group of bodonids: *Rynchomonas, Trypanoplasma*, and *Cryptobia helicis* (but, interestingly, not other species of *Cryptobia*), differ from the rest in having no transition-zone end-plates. These organisms also have their own unique version of an axosome; the small flat ring is replaced by a long cylinder (Brugerolle *et al.*, 1979; Burzell, 1973; Swale, 1973 (Figure 6).

The euglenoids may or may not have some sort of axosome. A granular platelike structure has been reported in *Euglena* (Piccinni and Mammi, 1978); a similar structure can be seen also in micrographs of *Astasia fennica* (see Figures 9, 11, in Fize and Michel, 1972) and a comparable, but larger, granular cylinder has been found in the posterior flagellum (only!) of *Entosiphon* (Mignot, 1964, 1966). In most micrographs of euglenoids, however, this region appears to be devoid of structure, the central tubules tapering away well distal to the transition zone (Figure 6).

In *Isonema* the central microtubules terminate at or near the distal plate in a manner not readily discernible in the published micrographs (Porter, 1973; Schuster *et al.*, 1968).

Two considerations of major systematic and phylogenetic significance may be

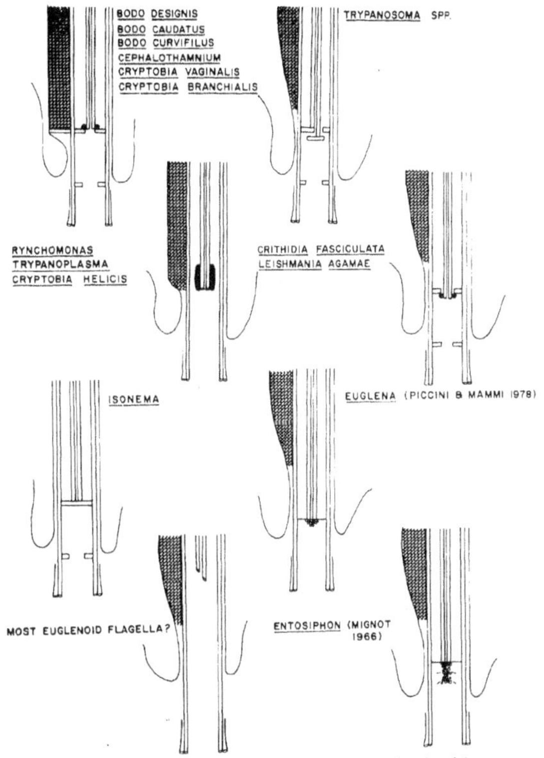

Fig. 6. Diagrammatic representations of flagellar transition zones of bodonids, trypanosomatids, *Isonema*, and euglenoids. Apart from the large paracrystalline paraflagellar rods (crosshatched), the specific structures depicted are described in the corresponding section of the text.

derived from this collective information. First, the euglenoids, the kinetoplastids, and *Isonema* are all members of a common phyletic super-group. Second, within this super-group an evolutionary sequence is suggested.

A common phyletic super-group. The available data strongly support the long-suspected close phyletic relationship of the euglenoids and the kinetoplastid flagellates.

The numerous morphological characteristics that we have surveyed in all these organisms can themselves be grouped into categories which correspond to the 'seme' terminology of Hanson (1976, 1977).

An overwhelming number of plesiosemes, (similar, or even identical, properties) such as the general morphology, contractile vacuole type, nuclear characteristics, inter-linked pellicular microtubules, and PFB (Figure 7), strongly indicate that the euglenoids, kinetoplastids, and *Isonema* form one common phylogenetic super-group. In the future, taxonomic treatments of these organisms will have to be revised to take this into account (as already anticipated by e.g., Corliss, 1981; Taylor, 1976).

Morphological similarities have also prompted previous evolutionary speculations: Leedale (1978) hypothesizes the trypanosomatids as ancestral to the euglenoids, while

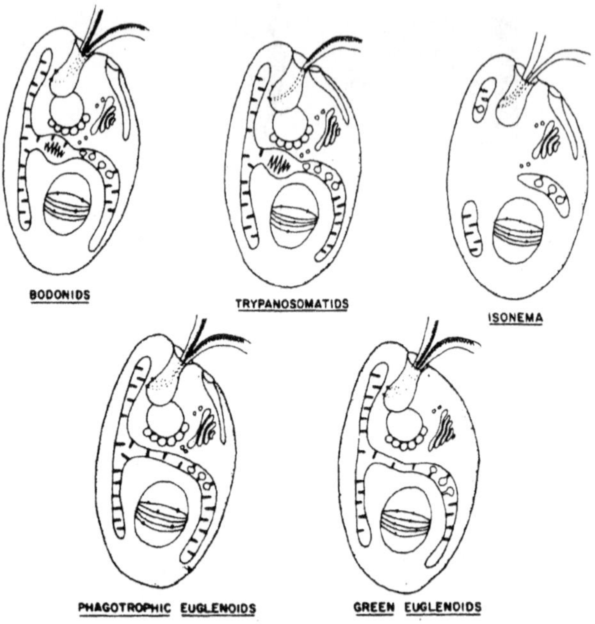

Fig. 7. *Plesiosemes*: Characteristics which most bodonids, trypanosomatids, *Isonema* and euglenoids have in common include external morphology, flagella inserted in a subapical pocket, paraflagellar rods, a tubular ingestion apparatus ventral to the flagellar pocket, a 'Type D' contractile vacuole associated with a Golgi apparatus, a single, giant mitochondrion with constricted-base flat cristae, and a completely closed mitosis.

Bovee (1972) derives the trypanosomatids from the euglenoids. We question both of these scenarios: as an entirely parasitic group, and therefore derived, the trypanosomatids are unlikely to be ancestral to anything else; moreover, their closest phyletic affiliations are clearly with the bodonids, as indicated by the amalgamation of the two groups as the Kinetoplastida (Brooker, 1971a; Honigberg, 1963).

Isonema, an evolutionary enigma, was initially taken to be a primitive euglenoid (Schuster *et al.*, 1968). Comparison of its characteristics with those of both the euglenoids and bodonids shows, however, that it could just as easily be categorized as an aberrant bodonid. The one particularly euglenoid characteristic of *Isonema* is the trichocyst. The batteries of trichocysts found in *I. nigricans* (Schuster *et al.*, 1968) resemble more closely those of *Entosiphon* (Mignot, 1966; Mignot and Hovasse, 1973) and *Peranema* (Hilenski and Walne, 1983) than they do the ejectile organelles of other groups of protists (e.g., Dodge, 1973) (Figure 1). Perhaps it is best to interpret this unusual creature as representing a bridge group affiliated with both the bodonids and euglenoids.

A likely evolutionary sequence. The cytoskeleton, ingestion apparatus support, and flagellar transition zone, however, are aposemes, (characteristics which, though variable, are recognizably homologous in the different taxonomic groups). These are particularly important because they allow a second point to be made, one not put forward previously: they suggest an evolutionary direction within this euglenoid-*Isonema*-kinetoplastid super-group.

Our interpretation of this group is as follows: within the assemblage, the green euglenoids are the most advanced forms; the euglenoid phagotrophs are less advanced, and the most primitive type is probably represented by *Bodo* and *Rynchomonas*. Quite likely the euglenoids and trypanosomatids both originated from advanced, *Cryptobia-like bodonids*, with *Isonema* being another offshoot of the bodonid-early euglenoid group (Figure 8).

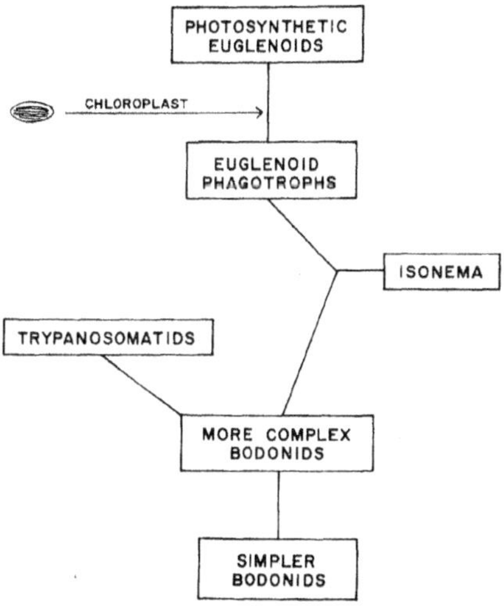

Fig. 8. An evolutionary pathway is postulated which links the euglenoids, bodonids, trypanosomatids and *Isonema*. In this scheme, phagotrophy is primitive and the introduction of chloroplasts and phototrophy is a late event.

This scheme would involve the progressive evolution of the cytoskeleton from separate microtubular bands, to a complete intracellular 'corset', to the addition of a pellicle consisting of a small number of relatively rigid strips to, finally a subordinate, less prominent, microtubular layer associated with a more advanced pellicle having numerous strips (Figure 9). Second, it would indicate a progressive increase in the complexity of the ingestion apparatus, including both the step-wise introduction of new structures and an increase in their size, followed by the eventual loss of the entire apparatus and its replacement by an autotrophic mode of nutrition. Third, the primitive flagella transition zone would be of the bodonid type, not too different from those of other protists (e.g., Dodge, 1973), evolving to the more distinctive types found in the trypanosomatids, in one direction, and in *Isonema* and euglenoids, in the other direction (Figure 9).

This sequence is, of course, in contrast to several other theories on the origin and development of these organisms, and conflicts perhaps most dramatically with the idea that the green euglenoids are ancestral to the colorless ones. This latter postulate has

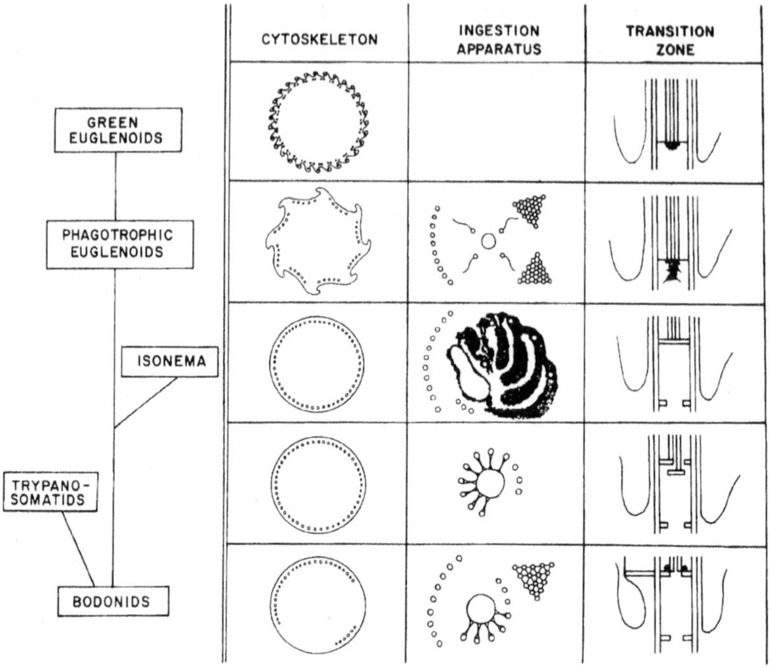

Fig. 9. *Aposemes*: Support for the postulated evolutionary pathway (left) is provided by these characteristics which vary from group to group but are recognizably homologous. In this scheme, the peripheral cytoskeleton becomes progressively more complex, as does the ingestion apparatus (followed by complete loss in the phototrophs), and the transition zone becomes more individualistic.

been widely held, in part because instances of spontaneous chloroplast loss were well known (Lwoff and Dusi, 1935; Pringsheim and Pringsheim, 1952; Ternitz, 1912), whereas mechanisms for chloroplast acquisition either were unknown or were (or perhaps still are, by some) considered unacceptable (e.g., Christen, 1962; Klein and Cronquist, 1967; Lwoff, 1951; Pringsheim, 1948). (The possibility of a non-phosynthetic origin for euglenoids was put forth by Jahn (1946) and later by Leedale (1978), who suggested phagotrophs as the early forms, with trypanosomes as their possible ancestors, and by Dodge (in Levandowsky, 1979), who presented a phylogenetic dendrogram with the euglenoids derived from 'trypanosomes, etc.'; but in none of these instances was evidence presented to support the speculation.)

The reasoning behind our proposed scheme is essentially a form of negative logic: in order to interpret the aposeme differences along lines that are in keeping with the more traditional view that euglenoids gave rise to other groups (see e.g., Bovee, 1972; Christen, 1962; Fott, 1974; Hollande, 1952; Hutner *et al.*, 1968; Pringsheim, 1963; Taylor, 1976), one would have to propose a degenerate evolution in which: a) the pellicle strips become fewer and larger and are finally lost, and then the peripheral microtubules also become fewer and eventually are restricted to isolated bands; b) the elaborate euglenoid ingestion apparatus suddenly appears in its present complexity from no pre-existing structure, and through time evolves into progressively smaller and simpler versions until it becomes a simple microtubule-bordered tube; and (c) the flagellar transition zone begins as the largely featureless euglenoid type, then evolves into the more usual type found in the bodonids and/or the somewhat unusual trypanosomatid type.

In contrast to the above, we believe that our proposal, in which structural complexity increases steadily through the progressive evolution of a series of phagotrophs, and culminates in the substitution of chloroplasts (e.g., via symbiotic acquisition of a photosynthetic prokaryote or the chloroplast of another eukaryote – see Gibbs, 1978) for an ingestion apparatus in the green euglenoid autotrophs, provides the most feasible evolutionary scheme consistent with the available data. This argument will be further supplemented by a more extensive treatment of this subject (Kivic and Walne, in preparation).

Implications concerning algal origins. Our postulated evolutionary sequence is significant in another respect. As indicated in Figures 8 and 9, the euglenoid-*Isonema*-kinetoplastid group represents a long line of phagotrophs into which a chloroplast was abruptly introduced to produce a series of non-phagotrophic photosynthetic algae. This is, of course, the sequence of events postulated for the origin of eukaryotic plants by the well-known Serial Endosymbiosis Theory (SET) (Margulis, 1968, 1970, 1981; Raven, 1970; Taylor, 1974, 1980), albeit reached in this case by an unrelated line of reasoning. Thus far, the SET has been discussed chiefly in hypothetical terms; however, here is perhaps the first example of a specific protistan evolutionary line in which the chloroplast acquisition event can be approximately placed. As a point of further interest, a number of organisms representing both the early photosynthetic forms and what might be taken to be their archaic pre-algal ancestors are still extant (thus also reinforcing the prediction of Round (1980) that the bodonids and trypanosomatids would be found to be pre-algal offshoots of the euglenoid-trypanosomatid-bodonid cluster). So, the SET version of chloroplast origins, presumed to have been a very ancient event, is now given further support by a concrete example, with the organisms that may represent the evidence still alive and available for further investigation.

Possible associations with other algal phyla. Finally, the wider affiliations of the euglenoid-*Isonema*-kinetoplastid group as a whole are another interesting problem. Despite their isolated position among the algae, a distant affiliation of the euglenoids with the chromophyte phyla – the haptophytes, chrysophytes and the dinoflagellates, in particular – has been suspected (e.g., Dodge, 1973; Leedale, 1967). Three particularly relevant examples can be cited which suggest such connections:

(a) Mignot (1967) notes that *Cyathomonas*, a cryptomonad, has an ingestion apparatus generally similar to that of the euglenoids and kinetoplastids.

(b) Chrysolaminarin (leucosin), the storage carbohydrate characteristic of the chrysophytes and haptophytes (Dodge, 1973), is chemically similar to paramylon, its counterpart in euglenoids. In some euglenoids, chrysolaminarin may be the main storage product (Leedale, 1967; Throndsen, 1973), whereas the haptophyte *Pavlova*

(Kreger and Van der Veer, 1979) and the cryptophyte *Chilomonas* (Hutchens *et al.*, 1948) are both reported to store paramylon. Perhaps this ability to synthesize straight-chain β1,3 glucans is itself a characteristic with phylogenetic significance.

(c) The PFR, a distinctive structure which has been shown here to be one of the common characteristics of the euglenoids and kinetoplastids, is present also in *Oxyrrhis* and a number of other dinoflagellates (Dodge, 1973; Dodge and Crawford, 1971; Moestrup, 1982). Further, although most dinoflagellates possess a mesokaryotic nucleus and peculiar form of mitosis characteristic of this group, *Oxyrrhis* appears to represent a sub-group in which both nuclear morphology and the mechanism of nuclear division are quite similar to those of euglenoids; the possibility of a phylogenetic relationship already has been suggested (Triemer, 1982).

Should such a connection actually exist, the conclusions of the present paper suggest that these algae are related most directly through their pre-chloroplastidic, phagotrophic, precursors (Figure 10).

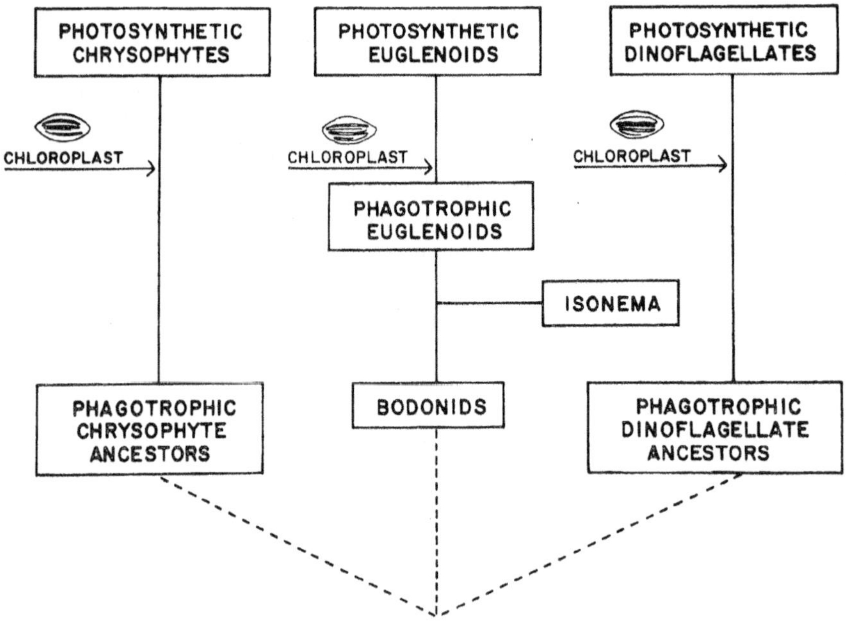

Fig. 10. Some evidence suggests a distant phylogenetic relationship of euglenophytes to certain other algal phyla. One possible interpretation of this is a common ancestry via their pre-algal precursors, pre-dating the acquisition of chloroplasts.

In summary, it is well established that the euglenoids have an isolated position among the algae; the reason for this is now clear. We suggest here that their closest relatives, and probable ancestors, are none of the algae, but the kinetoplastid zooflagellates. The postulated origin of the euglenoid autotrophs from a long line of phagotrophs via the relatively recent introduction of a chloroplast fits well with the

SET and the hypothesized origins of plant cells via the endosymbiosis of chloroplasts. This may be the first concrete example in support of these speculations. We anticipate that equivalent pre-algal precursors may be found for other algal groups, and speculate that the euglenoids and other phytoflagellates may even share common ancestry by way of these phagotrophic ancestors.

Acknowledgements

This work was supported in part by NSF grant 8208401 to P. L. W. We are grateful to Ms. Liz Shugart for the diagrams and to Mr. John Dunlap for photographic assistance.

References

Anderson, W. A. and Ellis, R. H.: 1965, Ultrastructure of *Trypanosoma lewisi*: Flagellum, Microtubules, and the Kinetoplast, *J. Protozool.* **12**, 483–499.

Angelopoulos, E.: 1970, Pellicular Microtubules in the Family Trypanosomatidae, *J. Protozool.* **17**, 39–51.

Bodammer, J. E. and Sawyer, T. K.: 1981, Aufwuchs Protozoa and Bacteria on the Gills of the Rock Crab, *Cancer irroratus* Say: A Survey by Light and Electron Microscopy, *J. Protozool.* **28**, 35–46.

Bovee, E.C.: 1972, The Trypanosomatidae: Possibly Descendant from Euglenoids? *J. Protozool.* Suppl. **18**, 14.

Brack, C.: 1968, Elektronmikroskopische Untersuchungen zum Lebenszyklus von *Trypanosoma cruzi*, *Acta Tropica* **25**, 289–351.

Brooker, B. E.: 1971a, Fine Structure of *Bodo saltans* and *Bodo caudatus* (Zoomastigophora: Protozoa), and their Affinities with the Trypanosomatidae, *Bull. Brit. Mus. Nat. History* **22**, 87–102.

Brooker, B. E.: 1971b, The Fine Structure of *Crithidia fasciculata* with Special Reference to the Organelles Involved in the Ingestion and Digestion of Protein, *Z. Zellforsch. Mikroskop. Anat.* **116**, 532–563.

Brooker, B. E. and Ogden, C. G.: 1972, Encystment of *Bodo caudatus*, *Protoplasma* **74**, 397–409.

Brown, K. N., Armstrong, J. A., and Valentine, R. C.: 1965, The Ingestion of Protein Molecules by Blood Forms of *Trypanosoma rhodesiense*, *Exp. Cell Res.* **39**, 129–135.

Brugerolle, G., Lom, J., Nohynkova, E., and Joyon, L.: 1979, Comparison et Evolution Cellulaires chez Pluseurs Especes de Bodonides et Cryptobiides Appartenant aux Genres *Bodo*, *Cryptobia*, et *Trypanoplasma* (Kinetoplastida, Mastigophora), *Protistologica* **15**, 197–221.

Burzell, L. A.: 1973, Observations on the Proboscis-Cytopharynx Complex and Flagella of *Rynchomonas metabolita* Pshenin 1964 (*Zoomastigophorea, Bodonidae*), *J. Protozool.* **20**, 385–393.

Burzell, L. A.: 1975, Fine Structure of *Bodo curvifilus* Griessmann (Kinetoplastida: Bodonidae), *J. Protozool.* **22**, 35–39.

Chen, Y. T.: 1950, Investigations on the Biology of *Peranema trichophorum* (Eugleninae), *Quart. J. Microscop. Sci.* **91**, 279–308.

Christen, H. R.: 1959, New Colorless Eugleninae, *J. Protozool.* **6**, 292–303.

Christen, H. R.: 1962, Zur Taxonomie der farblosen Euglenien, *Nova Hedwigia* **4**, 437-464.

Clark, T. B.: 1959, Comparative Morphology of Four Genera of Trypanosomatidae, *J. Protozool.* **6**, 227–232.

Corliss, J. O.: 1981, What Are the Taxonomic and Evolutionary Relationships of the Protozoa to the Protista?, *BioSystems* **14**, 445–459.

DeSouza, W., Rossi, M. A., Kitajima, E. W., Santos, R. R., and Roitman, I.: 1976, An Electron Microscopic Study of *Herpetomonas* sp. and *Leptomonas pessaoi*, *Can. J. Microbiol.* **22**, 197–203.

Dodge, J. D.: 1973, *The Fine Structure of Algal Cells* (Academic Press, London and New York) 261 p.

Dodge, J. D.: 1975, The Fine Structure of *Trachelomonas* (Euglenophyceae), *Arch. Protistenk.* **117**, 65–77.

Dodge, J. D. and Crawford, R. M.: 1971, Fine structure of the dinoflagellate *Oxyrrhis marina*. II. The Flagellar System, *Protistologica* **7**, 399–409.

Dynesius, R. A. and Walne, P. L.: 1975, Ultrastructure of the Reservoir and Flagella in *Phacus pleuronectes* (Euglenophyceae), *J. Phycol.* **11**, 125–130.

Eyden, B. P.: 1977, Morphology and Ultrastructure of *Bodo designis* Skuja 1948, *Protistologica* **13**, 169–179.

Fize, A. and Michel, R.: 1972, Ultrastructure d'*Astasia fennica* Michajlow (Euglenida) Parasite du Tube Digesif de Copepodes *Cyclopidae*, *Protistologica* **8**, 143–153.

Freymuller, E. and Camargo, E. P.: 1981, Ultrastructural Differences between Species of Trypanosomatids with and without Endosymbionts, *J. Protozool.* **28**, 175–182.

Fuge, H.: 1969, Electron Microscope Studies on the Intra-flagellar Structures of Trypanosomes, *J. Protozool.* **16**, 460–466.

Gardner, P. J.: 1974, Pellicle-associated Structures in the Amastigote Stage of *Trypanosoma cruzi* and *Leishmania* species, *Ann. Trop. Med. Parasit.* **68**, 167–176.

Grasse, P. P.: 1952, Ordre des Trypanosomides, in P. P. Grasse, (ed.), *Traite de Zoologie, Vol. 1*, (Masson et cie., Paris) pp. 602–668.

Hanson, E. D.: 1976, Major Evolutionary Trends in Animal Protists, *J. Protozool.* **23**, 4–12.

Hanson, E. D.: 1977, The Origin and Early Evolution of Animals (Wesleyan University Press, Middletown) 670 p.

Hilenski, L. L. and Walne, P. L.: 1983, Ultrastructure of Ejectile Mucocysts in *Peranema trichophorum* (Euglenophyceae), *J. Protozool.* **30**, 491–496.

Hitchen, E. T.: 1974, The Fine Structure of the Colonial Kinetoplast Flagellate *Cephalothamnium cyclopum* Stein, *J. Protozool.* **21**, 221–231.

Hollande, A.: 1952a, Classe des Eugleniens, in P. P. Grasse (ed.), *Traite de Zoologie, Vol. 1*, (Masson et cie., Paris) pp. 238–284.

Hollande, A.: 1952b, Ordre des Bodonides, in P. P. Grasse (ed.), *Traite de Zoologie, Vol. 1*, (Masson et cie., Paris) pp. 669–693.

Honigberg, B. M.: 1963, A Contribution to Systematics of The Non-pigmented Flagellates, in J. Ludvik, J. Lom, J. Vavra, (eds.), *Progress in Protozoology*, (Academic Press, London and New York) pp. 68–69.

Hutchens, J. O., Podolsky, B., and Morales, M. F.: 1948, Studies on the Kinetics and Energetics of Carbon and Nitrogen Metabolism of *Chilomonas paramecium*, *J. Cell. Comp. Physiol.* **32**, 117–141.

Hutner, S. H., Zahalsky, A. C., and Aaronson, S.: 1968, Inhibitors of Growth and Metabolism, in D. E. Buetow (ed.), *The Biology of Euglena, Vol. II*, Academic Press, New York, pp. 193–214.

Hyams, J.: 1982, The *Euglena* Paraflagellar Rod: Structure, Relationship to other Flagellar Components, and Preliminary Biochemical Characterization, *J. Cell Sci.* **55**, 199–210.

Joyon, L. and Lom, J.: 1966, Sur l'Ultrastructure de *Costia necatrix* LeClerq (Zooflagelle); Place Systematique de ce Protiste, *Comp. Rend. Acad. Sci. Paris* **262**, 660–663.

Joyon, L. and Lom, J.: 1969, Etude Cytologique, Systematique, et Pathologique *d'Icthybodo necator* (Henneguy 1883) Pinto 1928 (Zooflagelle), *J. Protozool.* **16**, 703–719.

Kivic, P. A. and Vesk, M.: 1972, Structure and Function in the Euglenoid Eyespot Apparatus: The Fine Structure, and Response to Environmental Changes, *Planta* (*Berl.*) **105**, 1–14.

Klein, R. M. and Cronquist, A.: 1967, A Consideration of the Evolutionary and Taxonomic Significance of Some Biochemical, Micromorphological and Physiological Characters in the Thallophytes, *Quart. Rev. Biol.* **42**, 105–296.

Kreger, D. R. and Van der Veer, J.: 1979, Paramylon in a Chrysophyte, *Acta Bot. Neerl.* **19**, 401–402.

Leedale, G. F.: 1967, *Euglenoid Flagellates*, Prentice-Hall, Englewood Cliffs, 242 p.

Leedale, G. F.: 1970, Observations on the Mitochondrial Reticulum in Living *Euglena gracilis*, *Cytobiologie* **1**, 195–202.

Leedale, G. F.: 1978, Phylogenetic Criteria in Euglenoid Flagellates, *BioSystems* **10**, 183–187.

Leedale, G. F. and Hibberd, D. J.: 1974, Obervations on the Cytology and Fine Structure of the Euglenoid Genera *Menoidium* Perty and *Rhabdomonas* Fresenius, *Arch. Protistenk.* **116**, 319–345.

Levandowsky, M.: 1979, Unity and Diversity in Biochemistry, *Quart. Rev. Biol.* **54**, 55–60.

Linder, J. C. and Staehlin, L. A.: 1977, Plasma Membrane Specializations in a Trypanosomatid Flagellate, *J. Ultrastr. Res.* **60**, 246–262.

Lom, J., Paulin, J. J., and Nohynkova, E.: 1980, The Fine Structure of the Fish Trypanosome *Trypanosoma danilewskyi*: 1. Presence of a Cytopharyngeal Complex in Bloodstream Trypomastigotes, *Protistologica* **16**, 365–373.

Lwoff, A.: 1951, Introduction to Biochemistry of the Protozoa, in A. Lwoff (ed.), *Biochemistry and Physiology of Protozoa*, I, (Academic Press, NY) pp. 1–28.

Lwoff, A. and Dusi, H.: 1935, La Suppression Experimentale des Chloroplastes chez *Euglena mesnili*, *Comp. rend. Soc. Biol. Paris* **119**, 1092–1095.

MacDonald, C. M. and Darbyshire, J. F.: 1977, The Morphology of a Soil Flagellate, *Spiromonas angusta* (Duj.) Alexeieff (Mastigophorea: Protozoa), *Protistologica* **13**, 441–450.
Margulis, L.: 1968, Evolutionary Criteria in Thallophytes: A Radical Alternative, *Science* **161**, 1020–1022.
Margulis, L.: 1970, *Origins of Eukaryotic Cells*, Yale University Press, New Haven, 349 p.
Margulis, L.: 1981, *Symbiosis in Cell Evolution*, W. H. Freeman and Co., San Francisco, 419 p.
McGhee, R. B. and Cosgrove, W. B.: 1980, Biology and Physiology of the Lower Trypanosomatidae, *Microbiol. Revs.* **44**, 140–173.
Meyer, H. and DeSouza, W.: 1973, On the Fine Structure of *Trypanosoma cruzi*, in Tissue Cultures of Pigment Epithelium From Chick Embryo. Uptake of Melanin Granules by the Parasite, *J. Protozool.* **20**, 590–593.
Mignot, J.-P.: 1963, Quelques Particularites de l'Ultrastructure *d'Entosiphon sulcatum* (Duj.) Stein, Flagelle, Euglenien, *Comp. rend. Acad. Sci. Paris* **257**, 2530–2533.
Mignot, J.-P.: 1964, Observations Complementaires sur la Structure des Flagelles *d'Entosiphon sulcatum* (Duj.) Stein, Flagelle Euglenien, *Comp. rend. Acad. Sci. Paris* **258**, 3360–3363.
Mignot, J.-P.: 1966, Structure et Ultrastructure de Quelques Euglenomonadines, *Protistologica* **2**, 51–117.
Mignot, J.-P.: 1967, Affinites des Euglenomonadines et des Chloromonadines; Remarques sur la Systematique des Euglenida, *Protistologica* **3**, 5–23.
Mignot, J.-P. and Hovasse, R.: 1973, Nouvelle Contribution a la Connaissance des Trichocystes: Les Organites Grillages *d'Entosiphon sulcatum* (Flagellata, Euglenida), *Protistologica* **9**, 373–391.
Mikolajczyk, E. and Kuznicki, L.: 1981, Body Contraction and Ultrastructure of *Euglena*, *Acta Protozool.* **20**, 1–24.
Moestrup, Ø.: 1982, Flagellar Structure in Algae: A Review, with New Observations on the Chrysophyceae, Phaeophyceae (Fucophyceae), Euglenophyceae, and *Reckertia*, *Phycologia*, **21**, 427–528.
Nisbet, B.: 1974, An Ultrastructural Study of the Feeding Habits of *Peranema trichophorum*, *J. Protozool.* **21**, 39–48.
Patterson, D. J.: 1980, Contractile Vacuoles and Associated Structures: Their Organization and Function, *Biol. Rev.* **55**, 1–46.
Paulin, J. J.: 1975, The Chondriome of Selected Trypanosomatids. A Three-dimensional Study Based on Serial Thick Sections and High Voltage Electron Microscopy, *J. Cell Biol.* **66**, 404–413.
Paulin, J. J.: 1977, *Crithidia fasciculata*. Reconstruction of the Mitochondrion Based on Serial Thick Sections and High Voltage Electron Microscopy, *Exp. Parasit.* **41**, 283–289.
Paulin, J. J. and McGhee, R. B.: 1971, An Ultrastructural Study of the Trypanosomatid *Phytomonas elmassiani* from the Milkweed *Asclepias syriaca*, *J. Parasitol.* **57**, 1279–1287.
Pellegrini, M.: 1980, Three-dimensional Reconstruction of Organelles in *Euglena gracilis* Z. I. Qualitative and Quantitive Changes of Chloroplasts and Mitochondrial Reticulum in Synchronous Photoautotrophic Culture, *J. Cell Sci.* **43**, 137–166.
Piccinni, E. and Mammi, M.: 1978, Motor Apparatus of *Euglena gracilis*: Ultrastructure of the Basal Portion of the Flagellum and the Paraflagellar Body, *Boll. Zool.* **45**, 405–444.
Pitelka, D. R.: 1963a, *Electron Microscopic Structure of Protozoa*, Pergamon Press, Oxford, 269 p.
Pitelka, D. R.: 1963b, The Fine Structure of *Bodo saltans*, in: J. Ludvik, J. Lom, J. Vavra (eds.), *Progress in Protozoology*, (Academic Press, London and New York) p. 405.
Porter, D.: 1973, *Isonema papillatum*, sp.n., A New Colorless Marine Flagellate: A Light- and Electron Microscopic Study, *J. Protozool.* **20**, 351–356.
Postell, F. J. and McGhee, R. B.: 1981, An Ultrastructural Study of *Phytomonas davidi* Lafont (Trypanosomatidae), *J. Protozool.* **28**, 78–83.
Preston, T. M.: 1969, The Form and Function of the Cytostome-Cytopharynx of the Culture Forms of the Elasmobranch Hemoflagellate *Trypanosoma raiae*, *J. Protozool.* **16**, 320–333.
Pringsheim, E. G.: 1948, Taxonomic Problems in the Euglenineae, *Biol. Rev.* (*Cambridge Phil. Soc.*) **23**, 46–61.
Pringsheim, E. G.: 1956, Contributions Towards a Monograph on the Genus *Euglena*, *Acta Nova Leopoldina* **125**, 1–168.
Pringsheim, E. G. and Pringsheim, O.: 1952, Experimental Elimination of Chromatophores and Eyespot in *Euglena gracilis*, *New Phytol.* **51**, 65–75.
Raikov, I. B.: 1978, *The Protozoan Nucleus: Morphology and Evolution*, Springer-Verlag, Wein and New York, (Russian-to-English translation: 1982) 474 p.
Raven, P.: 1970, A Multiple Origin of Plastids and Mitochondria, *Science* **169**, 641–646.

Rosowski, J. R. and Lee, K. W.: 1978, *Cryptoglena pigra*: A Euglenoid with One Chloroplast, *J. Phycol.* **14**, 160–166.
Round, F. E.: 1980, The Evolution of Pigmented and Unpigmented Unicells – A Reconsideration of the Protista, *BioSystems* **12**, 61–70.
Schubert, G.: 1966, Zur Ultracytobiologie von *Costa necatrix* Le Clerq unter besonderes Berucksichtigung des Kinetoplast Mitochondrions, *Z. Parasitenk.* **27**, 271–286.
Schuster, F. L. and Hershenov, B.: 1974, *Kwakinea quartana*, A Colorless Euglenoid Flagellate. I. Ultrastructure, *J. Protozool.* **21**, 33–39.
Schuster, F. L., Goldstein, S., and Hershenov, B.: 1968, Ultrastructure of a Flagellate, *Isonema nigricans* nov. gen. nov. sp., from a Polluted Marine Habitat, *Protistologica* **4**, 141–149.
Steinert, M. and Novikoff, A. B.: 1960, The Existence of a Cytostome and the Occurrence of Pinocytosis in the Trypanosome *Trypanosoma mega*, *J. Biochem. Cytol.* **8**, 563–569.
Stewart, K. D. and Mattox, K. R.: 1980, Phylogeny of Phytoflagellates, in E. R. Cox (ed.), *Phytoflagellates*, Vol. II; Developments in Marine Biology, Elsevier/North Holland, New York, pp. 433–462.
Swale, E. M. F.: 1973, A Study of the Colorless Flagellate *Rynchomonas nastuta* (Stockes) Klebs, *Biol. J. Linn. Soc.* **5**, 255–264.
Taylor, F. J. R.: 1974, Implications and Extensions of the Serial Endosymbiosis Theory and the Origin of Eukaryotes, *Taxon* **23**, 229–258.
Taylor, F. J. R.: 1976, Flagellate Phylogeny: A Study in Conflicts, *J Protozool.* **23**, 28–40.
Taylor, F. J. R.: 1980, The Stimulation of Cell Research by Endosymbiotic Hypotheses for the Origin of the Eukaryotes, in W. Schwemmer and H. E. A. Schenk (eds.), *Endocytobiology*, Walter de Gruyter, Berlin and New York, pp. 917–947.
Ternitz, C.: 1912, Beitrage zur Morphologie und Physiologie der *Euglena gracilis*, *Jahrb. Wiss. Bot.* **51**, 435–514.
Throndsen, J.: 1973, Fine Structure of *Eutreptiella marina* (Euglenophyceae), *Norwegian J. Bot.* **20**, 271–280.
Triemer, R. E.: 1982, A Unique Mitotic Variation in the Marine Dinoflagellate *Oxyrrhis marina* (Pyrrophyta), *J. Phycol.* **18**, 399–411.
Vickerman, K.: 1963, The Fine Structure of *Blastocrithidia familiaris* (Gibbs) in Comparison with Other Trypanosomatid Flagellates, in J. Ludvik, J. Lom, J. Vavra (eds.), *Progress in Protozoology*, Academic Press, London and New York, pp. 398–401.
Vickerman, K.: 1971, Morphological and Physiological Considerations of Extracellular Blood Protozoa, in A. M. Fallis (ed.), *Ecology and Physiology of Parasites*, University of Toronto Press, Toronto, pp. 58–91.
Vickerman, K.: 1977, DNA Throughout the Single Mitochondrion of a Kinetoplastid Flagellate: Observations on the Ultrastructure of *Cryptobia vaginalis* (Hesse 1910), *J. Protozool.* **24**, 221–233.
Vickerman, K. and Preston, T. M.: 1976, Comparative Cell Biology of the Kinetoplastid Flagellates, in W. H. R. Lumsden and D. A. Evans (eds.), *Biology of the Kinetoplastida, Vol. I.*, Academic Press, London, pp. 35–130.
West, L. K., Walne, P. L., and Rosowski, J.: 1980, *Trachelomonas hispida* var. *coronata* (Euglenophyceae). I. Ultrastructure of Cytoskeletal and Flagellar Systems, *J. Phycol.* **16**, 489–497.
Willey, R. L., Durban, E. M., and Bowen, W. R.: 1973, Ultrastructural Observations of a *Colacium* Palmella: The Reservoir, Eyespot, and Flagella, *J. Phycol.* **9**, 211–215.

A GREEN ALGA WITH MINIMAL EUKARYOTIC FEATURES: *NANOCHLORUM EUCARYOTUM*

RUDOLF K. ZAHN

Academy of Science and Literature, Mainz; Institute of Physiological Chemistry of the University, D-6500 Mainz, F.R.G.
and
Laboratory for Marine Molecular Biology, Center for Marine Research, Y-52210 Rovinj, Yugoslavia, Rudjer Boskovic Institute, Zagreb, Yugoslavia

(Received 12 December, 1983)

1. Introduction

Recently we discovered a new, very small eukaryotic alga *Nanochlorum eucaryotum* (*N.e.*) (Wilhelm *et al.*, 1982) in our sea water aquarium in the institute at Mainz. Growth and photosynthesis of *Nanochlorum* have been described (Wilhelm and Wild, 1982). The organism grows autotrophically in our laboratory in continuous culture in logarithmic phase at more than twice the salt concentration of sea water with a generation time of 27 hr.

The known properties of *N. e.* so far reported are given in Table I. Strangely, for more than 300 generations meiosis or mitosis never have been observed. Before division starts the nucleus enlarges without any visible change in the appearence of the nuclear content, the inner leaflet of the nuclear double membrane grows a circular fold, enlarging towards the center and finally pinching the otherwise unaltered nuclear content into halves.

Prior to nuclear fission however, the chloroplast and the single mitochondrion divide, a Golgi apparatus appears which doubles, and while the cell membrane invaginates, new wall material is formed, resulting in a doubled cell wall between the daughter cells, and finally divides the whole into two entities.

These findings raise some principle questions concerning the eukaryotic status of *N. e.* which has been discussed in the literature under different aspects (Margulis, 1981). In comparison to most 'fully eukaryotic' eukaryotes, *N. e.* shows deviations (Table II) and the question may be asked: What are the adequate and sufficient conditions respectively and their combinations characterizing the eukaryotic status?

We are aware of our actual unability to answer this question.

In this paper we show that *N. e.* apparently does not contain tubulin-like proteins that are susceptible to common mitotic spindle inhibitors, that it does not contain histones in concentrations found in most eukaryotes and that nucleosome-like features are absent from all of its subcellular particles.

The investigation of *N. e.* however may disclose in a model like fashion what the state

Origins of Life **13** (1984) 289-303. 0302-1688/84/0133-0289$02.25.

of a 'marginal eukaryote' is and how far deviations can go without disabling biological fitness.

In this context it seems of importance to report here of an organism resembling *N. e.* in all respects that does not stem from an artifical biotope since it has been isolated from a natural biotope in the Northern Adriatic.

TABLE I

Some characteristics of *Nanochlorum eucaryotum*

(1)	Cell volume:	$1.5 \pm 0.6\,\mu^3$ (under conditions of unbalanced growth: $4.1 \pm 0.6\,\mu^3$);
(2)	Cell wall:	10–30 nm thick, even, smooth;
(3)	Subcellular elements:	1 dictyosome;
		1 mitochondrion ($12 \pm 5\%$ of the cell volume), spherical or ellipsoidal;
		1 chloroplast ($60 \pm 10\%$ of the cell volume), with chlorophyll a, b and with smooth bands of three to fivefold thylakoid lamellae, with plastoglobuli, without pyrenoids;
		1 nucleus ($21 \pm 5\%$ of the cell volume), with variable number of nucleopores;
		cytoplasm ($8 \pm 6\%$ of the cell volume, may increase considerably, e.g. with unbalanced growth), with few tubuli, that are covered with 16 ($\pm$ 4) nm thick particles;
(4)	Shape:	spherical, rarely slightly oval, sometimes double forms, otherwise single;
(5)	Nuclear division:	nuclear pores disappear; the nuclear membrane invaginates and pinches the nuclear content into two equal parts; the appearance of the nuclear content doesn't show alterations;
(6)	DNA content:	$6.1 \pm 0.8 \cdot 10^{-14}$ g/cell;
(7)	RNA content:	$2 \pm 1 \cdot 10^{-13}$ g/cell;
(8)	Protein content:	very variable , depending on growth conditions: $1 \cdot 10^{-13} - 1 \cdot 10^{-12}$ g/cell;
(9)	Energy requirement:	cell division and growth: 100–2500 lux; inhibition: 10^4 lx;
(10)	pH tolerance for growth:	pH 3–9;
(11)	Salt tolerance:	0–12% NaCl equivalents;

TABLE II

Features in *N. e.* deviating from typical eukaryotic state

1.	Size and DNA content are at the lowest ones found in typical eukaryotes.
2.	These single-celled organisms only contain of one single subcellular organelle each: chloroplast, mitochondrion, dictyosome.
3.	Nuclear division occurs without formation of condensed chromosomes.
4.	Only binary fission has been observed.
5.	Spindle apparatus, phragmoplast, fibrillar structures within the nucleus or outside of it never have been observed.
6.	The nuclear membrane never dissolves or even loosens up.

2. Material and Methods

2.1. Search for *N. e.* in natural biotopes

38 samples were taken in the Northern Adriatic from the shores up to 5 nautical miles from the Center for Marine Research at Rovinj (Istria, Yugoslavia) at all the places where we during the previous 2 years had taken samples, since ultimately from one of them the *N. e.* detected in our sea water aquarium in Mainz should stem.

13 samples were water samples from the surface and 13 from the bottom at 24–32 m. They were filtered by gravity through a 3 µm filter. The filtrate was then put through a 25 mm .42 µm filter (Millipore, Neu-Isenburg, F.R.G.). These filters were cut into quarters and put into sterile flasks. To two of them, 10 ml of sterile Adria water was added, to the two others 10 ml of sterile solution according to Table III. They were plugged and put into a roller apparatus (30 rpm) under a 12 hr 2000 lx daylight, 12 hr dark cycle, at 22 °C.

9 samples were rinsed off from macro-algae, shells and sponges with sterile Adriatic water with 250 ml each.

3 samples came from stones. One of them, from 'Colona', the source of the successful preparation, was in a fine crack going 4 cm deep into an otherwise indisturbed, about 12 kg, rock of white calcium carbonate. When a piece had been knocked off with a hammer a 3 cm broad coloured band became visible (Figure 1).

The blue green zone showed single-celled organisms considerably larger than *Nanochlorum* without nuclear structures under the microscope, whereas in the bright green area among others, small green organisms of the expected diameter of around 1 µm could be discerned. The organisms of the different layers could be brushed off with sterilized instruments. They were suspended in 250 ml sterile artificial or natural sea water, that had been replenished either by 'terra rossa' extract (vide infra) or by a trace

TABLE III

Composition of the artificial culture medium used for isolation and propagation of *Nanochlorum e.*

No Component	mole/l	Concentration factor compared to sea water	*No* Component	mole/l	Concentration factor compared to sea water
1 NaCl	1.1	2.3	25 $NiSO_4 \cdot 6H_2O$	$5 \cdot 10^{-7}$	4.2
2 $MgCl_2$	$1.5 \cdot 10^{-1}$	2.8	26 $BaCl_2$	$5 \cdot 10^{-7}$	3.3
3 Na_2SO_4	$8.8 \cdot 10^{-2}$	3.1	27 KI	$5 \cdot 10^{-7}$	1.1
4 KNO_3	$3.3 \cdot 10^{-2}$	6.9	28 SeO_2	$1 \cdot 10^{-7}$	$9.1 \cdot 10$
5 KCl	$2.2 \cdot 10^{-2}$	2.2	29 $TiOSO_4 \cdot H_2O$	$1 \cdot 10^{-7}$	5.0
6 $CaCl_2$	$2.2 \cdot 10^{-2}$	2.2	30 $ZrOCl_2 \cdot 8H_2O$	$1 \cdot 10^{-7}$	3.0
7 KH_2PO_4	$1.1 \cdot 10^{-3}$	$3.8 \cdot 10^2$	31 As_2O_2	$1 \cdot 10^{-7}$	2.9
8 $NaHCO_3$	$1.1 \cdot 10^{-3}$	2.4	32 V_2O_5	$1 \cdot 10^{-7}$	2.6
9 H_3BO_3	$1.0 \cdot 10^{-3}$	2.4	33 $Ce(SO_4)_2 \cdot 4H_2O$	$5 \cdot 10^{-8}$	$7.1 \cdot 10^3$
10 KBr	$8 \cdot 10^{-4}$	0.95	34 $Y(NO_3)_3$	$5 \cdot 10^{-8}$	$4.6 \cdot 10^3$
11 LiCl	$5.0 \cdot 10^{-4}$	$2.0 \cdot 10$	35 $La(NO_3)_3$	$5 \cdot 10^{-8}$	$2.3 \cdot 10^3$
12 $Sr(NO_3)_2$	$5.0 \cdot 10^{-4}$	5.6	36 $Be(NO_3)_2 \cdot 4H_2O$	$5 \cdot 10^{-8}$	$7.1 \cdot 10^2$
13 Humic Subst.	$1.0 \cdot 10^{-4}$	$1.5 \cdot 10^2$	37 $Sc(NO_3)_3$	$5 \cdot 10^{-8}$	$5.6 \cdot 10^2$
14 NaF	$1.0 \cdot 10^{-4}$	1.5	38 $BiONO_3$	$5 \cdot 10^{-8}$	$5.0 \cdot 10^2$
15 Rb_2SO_4	$5.0 \cdot 10^{-5}$	$3.6 \cdot 10$	39 $TlNO_3$	$5 \cdot 10^{-8}$	$1.0 \cdot 10^2$
16 EDTA	$5 \cdot 10^{-5}$	—	40 $WO_3 \cdot H_2O$	$5 \cdot 10^{-8}$	$9.3 \cdot 10$
17 $FeSO_4$	$1.5 \cdot 10^{-5}$	$2.8 \cdot 10^2$	41 $HgCl_2$	$5 \cdot 10^{-8}$	$5.0 \cdot 10$
18 $MnCl_2 \cdot 4H_2O$	$5 \cdot 10^{-6}$	$1.4 \cdot 10^2$	42 PbOAc	$5 \cdot 10^{-8}$	$3.4 \cdot 10$
19 Ga_2O_3	$5 \cdot 10^{-6}$	1.2	43 CsCl	$5 \cdot 10^{-8}$	$2.5 \cdot 10$
20 $ZnSO_4$	$1 \cdot 10^{-6}$	6.7	44 $CdSO_4$	$5 \cdot 10^{-8}$	$2.2 \cdot 10$
21 MoO_3	$1 \cdot 10^{-6}$	$1 \cdot 10$	45 $SbCl_3$	$5 \cdot 10^{-8}$	$2.0 \cdot 10$
22 $Al_2(SO_4)_3$	$5 \cdot 10^{-7}$	$1.3 \cdot 10$	46 $AgNO_3$	$5 \cdot 10^{-8}$	$1.8 \cdot 10$
23 $Co(NO_3)_2 \cdot 6H_2O$	$5 \cdot 10^{-7}$	$7.4 \cdot 10$	47 UO_2Ac	$5 \cdot 10^{-8}$	3.9
24 $CuSO_4 \cdot 5H_2O$	$5 \cdot 10^{-7}$	$1.1 \cdot 10$	48 $K\,TeO(OH)_5H_2O$	$5 \cdot 10^{-8}$	–

element mixture that approximated its elementary composition to yield the solution of Table III.

The 250 ml suspension, in 500 ml Erlenmeyer flasks were shaken with 2 strokes per second, at an illumination of 800 lx, 22 °C for 6 weeks.

2.2. Preparation of Culture Fluid

The mixture in its final form (Table III) was composed of several stock solutions. Stock a: components 1–5, Stock b: component 6, Stock c: 7–12. Stock d: component 13, for which 1 kg of dry peat moss (Floratorf) + 30 g Kieselgel 'Silica' (Woelm, Eschwege, F.R.G.) were suspended in 3 l of distilled water and kept at 37 °C for 24 hr. It was then centrifuged and to the clear supernatant ethylene-diamine-tetraacetate-disodium (EDTA) was added. It was divided into two portions. In one of them the components 26–48 were dissolved respectively, the components, 14, 15 and 17–24 in the other, each one at 1000 fold concentration of the final culture medium. Component 25 was added to the final solution separately. The culture medium in its final composition is heated in

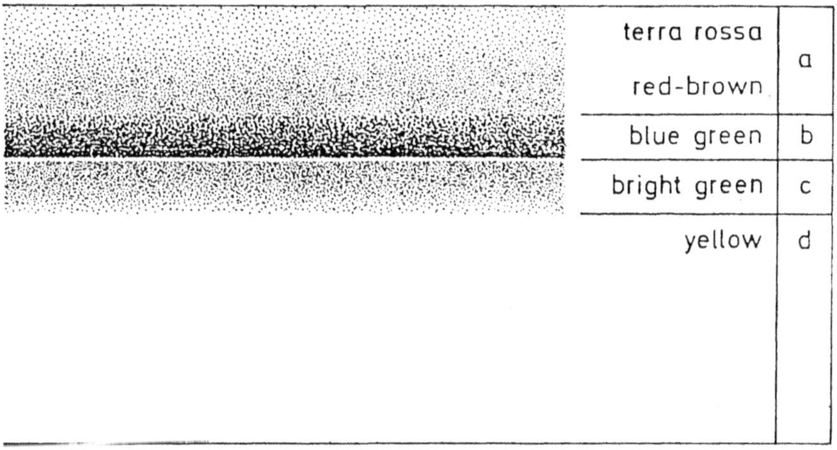

Fig. 1. Location of *Nanochlorum eucaryotum*. The sketch gives a view onto the wall of a crack of calcium carbonate rock from Colona. The upper heavy line corresponds to the rock's surface, from there on spreads a 1 cm wide red brown zone (a) coloured by fine soil particles (terra rossa) followed by a dark blue green band (b) inhabitated by cyanobacteria. Below this follows a bright green, 4 mm wide band, from which *Nanochlorum e.* could be isolated. Below this follows a faint yellow band, about 1.3 cm wide and then fresh fracture faces, emerge.

an autoclave to 110 °C for 30 min. On cooling turbidity arises, which however soon disappears when inoculated with more than 10^4 cells per ml.

For the first successful isolation experiments a 'terra rossa' extract has been used to supplement the artificial sea water made of the components 1–11, 13, and 14. For this purpose 1 kg of dried deposit from the mouth of a spring time fresh water well was extracted with 1 l of 6 M HCl in a roller bottle over night at 25 °C. It was centrifuged until clear and the supernatant evaporated under vacuum at 50 °C until a dry residue was formed. This was suspended as a fine slurry in 1 l of solution of components 12 and 15 and kept on a roller for a day, while it almost dissolved. Upon standing over night the supernatant was decanted and used instead of components 17–46 in a concentration of 1 ml per 1 l of solution with components 1–11, 13, and 14.

2.3. Continuous culture

5 l flasks with conical bottoms containing 4 l of solution (Table III) were inoculated with the content of a test tube, containing a cell clone and gassed with compressed air at 400 l per hr at 22 °C. The vessel was illuminated at 1000 lx and sterile solution (Table III) was pumped continuously into the 5 l flask at 2 l per day in such a way that the volume surpassing 4 l is driven out into a receiving bottle and harvested.

Whenever large amounts of cells are needed, air gassing is avoided and the cultures are gassed with natural gas from the city line (Table IV). Under these conditions the cells grow at their highest doubling rate and density. Further implications of this are still under investigation.

TABLE IV

Analysis of the natural gas used[a]

	Volume per cent
CH_4	84.70
N_2	5.60
C_2H_6	5.50
CO_2	1.86
C_3H_8	1.66
C_4H_{10}	0.53
C_5H_{12}	0.10
C_6H_{14}	0.03
C_7H_{16}	0.01
C_8H_{18}	0.01
C_6H_6	0.01

[a] We are grateful to the Stadtwerke Mainz AG for their analysis.

2.4. Immunological identification

For this white, male NMRI mice were injected with 50 mg or about 10^{10} of *Nanochlorum* cells, washed in 0.2 ml of saline solution.

Ten days later the mice were killed, their spleens removed, and the immunocytes liberated by gentle homogenization. The putative *Nanochlorum* cells, were added in some assays, and authentical *Nanochlorum eucaryotum (Mainz II)* cells, or *Chlorella* cells in others. After 2 hr incubation at 4 °C the rosettes were counted at a multiplicity of 1 spleen cell each for more than 10 *Nanochlorum* cells (Nowotny, 1979). Using this method *Nanochlorum* can be distinguished from other algae, e.g. different *Chlorella* (Deutsche Algensammlung, Göttingen, F.R.G.) species, even in small volumes.

2.5. Inhibitor assays

In these experiments cells were grown in 10 ml cultures in sets of ten parallel samples in roller tubes, at 30 rpm, 1000 lx, 22 °C in 12 hr light, 12 hr dark cycles with daily changes of culture fluid. Cell numbers were evaluated as turbidity difference of the growing culture and its supernatant after centrifugation at 1000 g for 10 min at 22 °C. The instrument was calibrated weekly for cell number correlation. Statistical significance evaluation was with the non-parametric X-test (Van der Waerden and Nievergelt, 1976).

Agents of poor water solubility were dissolved in 10 µl of dimethylsulfoxide (DMSO), which then was added to the 10 ml culture, the control receiving the additive without the inhibitor. The solutions were freshly prepared.

The experiments lasted up to 96 hr with positive values of generation times, otherwise they were stopped at 48 hr.

The cells were inspected by light microscopy after 48 hr for crude morphological changes, for contamination with alien organisms, and for precipitation.

Whenever feasible the highest possible concentration has been tested, as far solubility and the price allowed for. The active concentrations were diluted and tested until the effect has disappeared.

For the determination of intracellular inhibitor concentration, inhibition of L5178Y cell propagation has been used.

At the end of the experiment the *N. e.* cells were sedimented at 10 000 g for 20 min at 3 °C, washed with 1 volume of culture fluid (Table III), the dry sediment was suspended in 1 volume of 50% aqueous DMSO and subjected to a series of 5 cycles of freeze-thawing on dry ice and warming to 40 °C. It was then centrifuged at 15 000 g for 40 min at 25 °C. The clear supernatant was added as a concentration series to 5 roller tubes each with 3000 L5178Y mouse lymphoma cells per ml growing in suspension culture in Fischer medium with 10% horse serum. They were run in parallel to a concentration series of inhibitor added to 50% DMSO extracts of control *N.e.* cells. After an incubation time of 72 hr the cell numbers and their volume distribution were evaluated in a Coulter counter with size distribution plotter (Coulter Electronics, Hialeah, Fla., U.S.A.).

This procedure allowed for determination of intracellular inhibitor concentrations which we judge to be exact within one order of magnitude. With colchicine additional methods were applied.

^{3}H-Colchicine, spec. act. 2.5 Ci/mmole (Amersham-Buchler, Braunschweig, F.R.G.) was used. At low colchicine concentrations the intracellular level was estimated by radioactivity measurements of chloroform extracts of *N. e.* cells, where quenching is more than 90%. At $8 \cdot 10^{-6}$ mole colchicine/l in the growth medium the drug is accumulated by a factor of at least of 10 in 48 hr.

With the higher concentration at $8 \cdot 10^{-3}$ mole/l after 48 hr of incubation the *N.e.* cells were ground and extracted into chloroform, evaporated to dryness and chromatographed on HPTLC Kieselgel 60, F 254 nm plates (E. Merck, Darmstadt, F.R.G.), using the solvent system ethylacetate-isopropanol-aqueous ammonia (25%) (50 : 30 : 2 by Vol). Colchicine under these conditions appears as a spot, strongly quenching the fluorescence caused by 254 nm excitation. The colchicine quenching spot stays clear of the *N.e.* pigments. Its quenching intensity was evaluated using a TLC scanner.

2.6. Analytical methods

2.6.1. For DNA isolation *N. e.* sediment was suspended and frozen into micropellets as described recently (Zahn, 1983). These are shaken with 2 volumes of 1 mm glass beads in a tissue disintegrator (Braun, Melsungen, F.R.G.) with CO_2-cooling. The DNA was isolated and treated further (Zahn, 1983).

2.6.2. Further analytical methods were as described (Wilhelm *et al.*, 1981; Zahn, 1983).

2.6.3. For nucleosomal preparations and DNA periodicities caused by nucleosome beading we followed the recipe of Bodansky *et al.* (1979).

2.6.4. *Histones.* For histone preparations different procedures have been followed: Sonnenbichler, 1975; Monahan and Hall, 1975; Johns, 1977; Bidney and Reek, 1977; Yamada *et al.*, 1978.

For their characterization we tried several procedures, described by Luck *et al.*, 1958; von Holt and Brandt, 1977; Spring and Cole, 1977; Kays *et al.*, 1979; Certa and von Ehrenstein, 1981 (among others).

2.7. Electron microscopy

2.7.1 Preparation for spreading was as described (Zahn, 1983). In some cases squash osmotic shock preparations of *N.e.* were achieved by treatment with an enzyme inhibitor mix: EDTA 2 g/l, phenylmethylsulfonylfluoride 800 mg/l, Na-tetrathionate 100 mg/l, Na-iodoacetamide 100 mg/l, Na-di-sulfite 100 mg/l, pH 8.

The cells were sedimented (1000 g, 10 min, 22 °C) and the sediment taken up with 3–10 volumes of inhibitor mix in some cases 1 : 100 diluted with water. Some cells were disrupted by osmotic shock, exposing the nuclear content for spreading on activated carbon film. After rotary shadowing with Pt/Pd at an angle of 7 degrees and/or contrasting with uranylacetate, the specimens were ready for inspection.

In some cases spreading was successful with salt-free cytochrome solution (Kleinschmidt and Zahn, 1959). Electronmicroscopic evaluation was with an Elmiskop 101 with image amplifier (Siemens, Berlin, F.R.G.).

2.7.2. *Freeze fracture.* We essentially followed the procedures as described by Moor (1964; 1981). For this, *N. e.* cells were pelleted in capped Eppendorf tubes, and the dry pellet was taken up with a minimum of water, so that droplets of 50 µl could be placed on gold supports. (No glycerol was used). This was then frozen by submersion in liquid Freon 22 (Balzers, Liechtenstein) and placed into the thermostatted manipulation head of a freeze fracture apparatus, equipped with an electron gun evaporator (Balzers). After cutting with the wedge at a specimen temperature of −100 °C the specimen was etched in generally for 1 min. This was then coated at an angle of 45° with 2 nm of platinum, followed by a coating at 90° with 15 nm of carbon under the control of a quartz oscillator (Balzers).After disintegration of the adherent biological material by concentrated sulfuric acid, the carbon replicas were inspected with an electron microscopie at magnifications of 10 000–30 000 times.

3. Results and Discussion

A brief summary of our actual knowledge on *Nanochlorum eucaryotum*, stemming from experiments reported in this paper is given in Figure 2 and Table V.

3.1. Identity

Figure 2 gives a picture of freeze fracture through *N.e.* towards the end of its logarithmic growth phase, which shows the organism in its 'fattest' aspect. The round cell is bordered by a thin cell wall, which is partially split showing a slit. The cytoplasm can be recognized as the continuous phase between three large round compartments, showing coarse granulation and vacuoles. A Golgi apparatus can be discerned. The cellular organelles are surrounded by double membrane systems that show at some parts of their circumference. The largest body is the chloroplast with thylakoid bands,

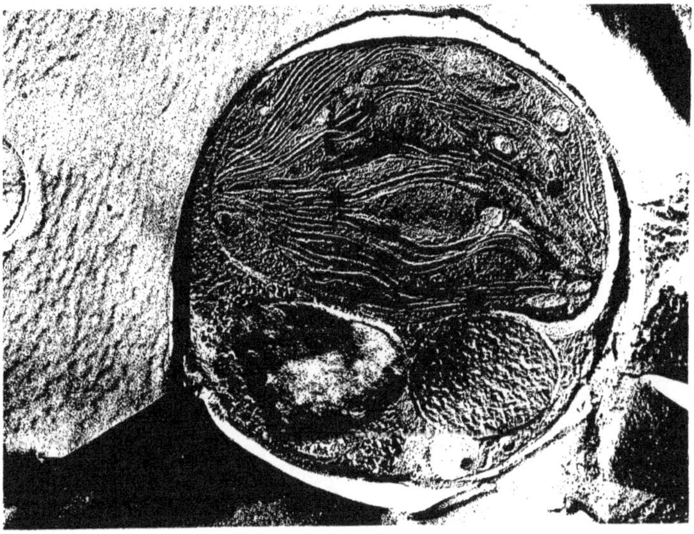

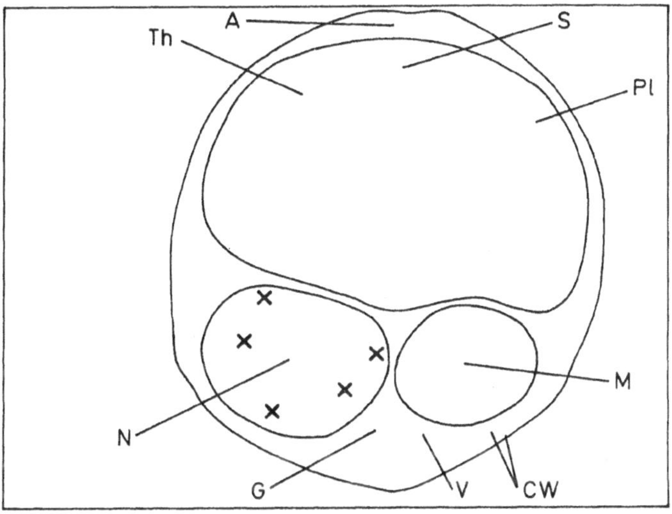

Fig. 2. Freeze fracture of *Nanochlorum eucaryotum* giving essential details of a cell of 1.4 µm × 1.6 µm. There is an artifactual gap (A) around the cell splitting the thin cell wall (CW) (30 nm). The largest subcellular body (about 60% of total cell volume (TCV)) is the plastid with thylakoid bands (Th). Between the thylakoids starch grana (S) and plastoglobuli (P1) are found in cells approaching the end of the logarithmic phase. The second largest body is the always nearly round nucleus (N, 21% TCV) with some nuclear pores (x), which partially has been stripped down to and across its nuclear membranes. The mitochondrion (M, 12% TCV) which suffered fracture across its inner structure, is always nearly round. The Golgi region (G), mostly with vacuoles (v), is sometimes visible, (Shadowing was from lower left corner).

TABLE V

Additional features of *Nanochlorum eucaryotum*

1.	Re-isolation from nature:	a morphologically, physiologically and immunologically similar organism;
2.	DNA base ratio:	$\frac{G + C}{(A + T) + (G + C)} \times 100 = 47.6 \pm 0.03\%$;
3.	5-methyl-cytosine content:	1 mole/100 mole bases;
4.	DNA superstructures:	only smooth and sometimes intertwined, filamentous structures have been isolated; nucleosomes are absent; no periodicities have been observed upon nuclease digestion;
5.	NO_3-reduction:	+
6.	Fe^{+++}-utilization:	+
7.	Inhibition of cell division:	(7.1) Colchicine[a,b] ($8 \cdot 10^{-6} - 8 \cdot 10^{-3}$ mole/l): no inhibition; (7.2) Colcemid ($8 \cdot 10^{-8} - 10^{-5}$ mole/l): no inhibition[b]; (7.3) Trifluralin ($10^{-5} - 3 \cdot 10^{-4}$ mole/l): no inhibition[b]; (7.4) Cytochalasin B ($2 \cdot 10^{-7} - 2 \cdot 10^{-5}$ mole/l): no inhibition[b]; (7.5) Azide ($1.5 \cdot 10^{-5} - 4.5 \cdot 10^{-4}$ mole/1): inhibition; (7.6) Diethylpyrocarbonate ($2 \cdot 10^{-6} - 6 \cdot 10^{-8}$ mole/l): inhibition; (7.7) Chloramphenicol ($3 \cdot 10^{-5} - 3 \cdot 10^{-3}$ mole/l): inhibition; (7.8) Cycloheximide (10^{-7}–10^{-4} mole/l): inhibition;

[a] Intracellular concentration determined by chromatography.
[b] Intracellular concentration determined by testing cell culture inhibition.

starch granules and plastoglobuli. A pyrenoid is lacking. The mitochondrion has fractured near its equator, exposing a coarse inner granulation. The etching procedure makes its envelope structure, and especially its close relationship to the chloroplast visible. This has been seen in all cases, whereas a close neighbourhood to the nucleus is not always observed. The nucleus, in most cases round, seems to be deformed. This is caused by incomplete ablation of surrounding material and of its outer membrane system. Five large nucleopores are visible.

The *N. e.* cells are tough, with their small radius of curvature hard to fracture. They are not staying in continuity with the embedding matrix.

From the morphological aspect the cells isolated from the Colona stone do not show any deviation from *N. e.* from Mainz. They have to be considered as morphologically identical.

The same holds for the immunological assay. The *N. e.* from Mainz and from Colona gave clear 'rosetting' (Nowotny, 1979) with relatively few large immunocytes, each one of them surrounded by a string of *N. e.* of more than 10 in one plane, while with *Chlorella pyrenoidosa*, *Chlorella fusca*, *Chlorella ellipsoidea* and *Chlorella vulgaris* no such aggregation can be seen. Some caution must be used however, when this technique is applied to cells as small as *N. e.* Since they do not adhere as strongly as larger animal cells do, vigorous shaking has to be avoided.

3.2. Histones

Since we suspected *N. e.* belongs to a group of organisms that never during their evolution had acquired all the features of full eukaryotism, we were interested in studying proteins which, throughout all taxa of eukaryotes are considered to be highly conserved i.e. histones (Isenberg, 1979). The enormous wealth of literature is rather scarce when it comes to small eukaryotes, the eukaryotic microbes (Horgen and Silver, 1978). Quite clearly, dinoflagellates like *Gymnodinium* do not have histones. Their basic nuclear, acid-extractable proteins differ substantially from histones; One consequence of this is that upon spreading of the DNA containing structures no 'beading' or DNA molecular weight clustering, typical for the higher eukaryotes, can be detected (Rizzo and Burghardt, 1982). In *Chlorophytes*, like *Chlorella ellipsoidea* and *Volvox*, the 'histones', in the form of acid-soluble proteins, proved to be quite different from calf-thymus histones. The same apparently holds for members of the order *Rhodymeniales* (Horgen and Silver, 1978).

This is perfectly in line with our intensive, fruitless searches for histones in *N. e.*. (Table V) Other eukaryotes considered to represent lower taxa, e.g. *Saccharomyces*, contain the highly conserved histones however (Isenberg, 1979). This absence of histones as in *N. e.* may thus become an important point in our understanding of eukaryotic evolution.

3.3. Inhibitors (Table V)

Azide was tested, since it is used as a sterilizing agent. It is a strong inhibitor of ATPase and a series of other enzymes at a range comparable to the one used for *N. e.* (Jain, 1982).

Diethylpyrocarbonate, another sterilizing agent, known to form amide bonds in proteins between free amino and carboxyl groups, is inhibitory in a range to be expected from the literature (Jain, 1982).

For the two drugs chloramphenicol and cycloheximide reciprocal inhibition was shown. The inhibition of protein synthesis by chloramphenicol in bacteria, 'blue-green algae', and in mitochondria and chloroplasts has been documented, and was absent in the cytoplasm of eukaryotes such as fungi, green algae, higher plants and mammals. The reverse has been reported to be true for cycloheximide. This is the consequence of their action on ribosomes, the chloramphenicol-sensitive 70 S procaryotic ribosomes being insensitive to cycloheximide, the 80 S cytoplasmic eukaryotic ribosomes being cycloheximide-sensitive and chloramphenicol-insensitive (Smillie and Scott, 1969; Jain, 1982).

In *N.e.* the amount of cytoplasm is relatively small, so that not many reserve substances and precursors seem to be present, making the macromolecular syntheses dependent on the metabolic status of the cellular organelles, especially of the large chloroplast. Thus is, besides a high sensitivity towards cycloheximide acting on the eukaryotic ribosomes in the cytoplasm, there is an appreciable inhibition by chloramphenicol, acting presumably on the 70 S ribosomes of the chloroplast (Table V). Thus while chloramphenicol is a 20–100 times better inhibitor of typical prokaryotic division at concentrations of $3 \cdot 10^{-6} - 3 \cdot 10^{-5}$ mole/l than on eukaryotes (Otten *et al.*, 1975) the higher value is overlapping with the minimal range in *N.e.* (Jain, 1982). In this context it will be interesting to find out, what the sensitivity will be under heterotrophic conditions.

Cytochalasin B, a drug that is known to inhibit intracellular motility phenomena (Dustin, 1978) at concentrations around or below $3 \cdot 10^{-5}$ mole/l (Jain, 1982) does not show measurable effects. Since determination of intracellular concentration may be inexact, not too much weight can be given to this observation however. For future experiments this introduces the interesting question of how an apparently marginal eukaryote solves its problems of shifting of subcellular structures.

3.4. Mitotic apparatus

This is considered as a leading marker in taxonomic evolution and has been incorporated as such into working hypotheses (e.g. Heath, 1980; Margulis, 1981). A frequently used poison for the synthesis of the mitotic apparatus and its function is colchicine. By reaction with one binding site (Bryan, 1972) of the tubulin subunits colchicine prevents their assembly into microtubules. This is the basis of its poisonous action, and leads to mitotic arrest (Dustin, 1978).

In *N. e.*, under conditions of apparently normal growth and propagation, colchicine is without significant effect, even at high concentrations. Therefore the question may be asked whether colchicine is able to enter the cells under the given conditions. At low concentrations in the culture fluid ($8 \cdot 10^{-6}$ mole/l) the drug is accumulated by a factor of at least 10 in the course of 48 hr, as measured with radiolabelled alkaloid. At the higher concentrations ($8 \cdot 10^{-2}$ mole/l) with the chromatographic method it could be shown that the outside concentration had been reached. Upon longer incubation periods, cells die, while thin flat, featherylike crystals appear occasionally at the outer cellular circumference, below the cell wall (Bauer and Zahn, unpublished). This affords based on the insensitivity of the 'mitotic' apparatus towards colchicine, colcemid and trifluralin (Table V) at the concentrations tested, and which had been found active in other taxa (Dustin, 1978; De Rosa *et al.*, 1978; Jain, 1982), this can signify that the colchicine binding site is not accessible in an otherwise functional protein or that such a protein may not be present. At the moment we cannot decide which of the two alternatives is present in *N. e.*. Investigations to clear up the matter are under way. As a working hypothesis we assume that a colchicine binding protein is present in *N.e.* giving rise to crystal formation. Whether such crystals are caused by a microtubular derivative which is not in the doublet form, corresponding to those of the marginal bundle of thrombocytes and of nucleated erythrocytes (Behnke and Forer, 1967; Behnke, 1970) is at the moment an open question.

We think it to be unlikely that a mitotic apparatus with typically active proteins like the tubulins of higher eukaryotes exists, since in such a case they ought to be inhibited by the colchicine treatment. As already shown in Table II and as substantiated by our further observations chromosome formation, dissolution of nuclear envelope, nor formation of fibrillar structures within, nor outside of the nucleus never have been observed. The deviations from the fully eukaryotic state in the mitotic apparatus and its function in *N.e.* establishes that this organism belongs to a taxon with marginal eukaryotic features, features that deserve further experimental efforts in order to be understood.

In this context the apparent absence of typical histones along with the missing nucleosome architecture may lead to the anticipation of possible alterations in genomic distribution and reshuffling in mitosis. Here the term 'mitosis' is to be considered in its wider sense (Dustin, 1978). This among other considerations raises the question about the fidelity of marker transmission in such organisms, e.g. if the replicated genomes 'attached to the nuclear envelope at a differentiated site" (Pickett-Heaps, 1974).

4. Conclusion

The possible total lack of histones, along with the apparent deviations in the protein properties and morphology of the mitotic apparatus indicate that *Nanochlorum eucaryotum* should be classified as a marginal eukaryote. An organism, closely resembling or identical with *N. e.* has been isolated from the Northern Adriatic Sea at Istria, Yugoslavia.

Investigations of marginal eukaryotic state with *N.e.* as model organism are expected to yield meaningful results with explanatory potential for our understanding of eukaryotic phylogeny.

Acknowledgements

I gratefully acknowledge the expert electronmicroscopic freeze fracture preparations of W. Bauer and V. Pondeljak, the skillful biochemical preparations of R. Beyer and E. Lamberth as well as the base analyses of Dr C. Schröder and Dr H.-J. Breter and the support by the Academy of Science and Literature, Mainz.

I want to thank my colleagues Prof. Dr A. Wild, Prof. Dr W. E. G. Müller, and Dr C. Wilhelm for stimulating discussions and Stuart Brown for aid in manuscript preparation.

References

Behnke, O.: 1970, 'Microtubules in Disk-Shaped Blood Cells', *Intern. Rev. Exp. Path.* **9**, 1–92.

Behnke, O. and Forer, A.: 1967, 'Evidence for Four Classes of Microtubules in Individual Cells', *J. Cell. Sci.* **2**, 169–192.

Bidney, D. L. and Reek, G. R.: 1977, 'Association Products and Conformations of Salt-Dissociated and Acid-Extracted Histones. A Two-Phase Procedure for Isolating Salt-Dissociated Histones', *Biochemistry* **16**, 1844–1849.

Bodansky, S., Mintz, L. B., and Holmes, D. S.: 1979, 'The Mesokaryote *Gyrodynium cohnii* Lacks Nucleosomes', *Biochem. Biophys. Res. Comm.* **88**, 1329–1336.

Borisy, G. G. and Taylor, E. W.: 1967, 'The Mechanism of Action of Colchicine. Binding of Colchicine-^{3}H to Cellular Protein', *J. Cell Biol.* **34**, 525–534.

Bryan, J.: 1972, 'Definition of Three Classes of Binding Sites in Isolated Microtubule Crystals', *Biochem.* **11**, 2611–2615.

Certa, U. and Von Ehrenstein, G.: 1981, 'Reversed-Phase High-Performance Liquid Chromatography of Histones', *Anal. Biochem.* **118**, 147–154.

DeRosa, F., Haber, D., Williams, C. and Margulis, L.: 1978, 'Inhibitory Effects of the Herbecide Trifluralin on the Establishment of the Cloverroot Module Symbiosis', *Cytobios* **21**, 37–43.

Dustin, P.: 1978, 'Microtubules', Spinger Verlag, Heidelberg, New York.

Heath, I. B.: 1980, 'Variant Mitoses in Lower Eukaryotes: Indicators of the Evolution of Mitosis', in G. H. Bourne, J. F. Danielli and K. W. Jean (eds), *International Review of Cytology*, Vol. 64, Academic Press, New York.

Holt, C. von and Brandt, W. F.: 1977, 'Fractionation of Histones on Molecular Sieve Matrices', *Methods in Cell Biology* **XVI**, 205–225.

Horgen, P. A. and Silver, J. C.: 1978, 'Chromatin in Eukaryotic Microbes', *Ann. Rev. Microbiol.* **32**, 241–284.

Isenberg, I.: 1979, 'Histones', *Ann. Rev. Biochem.* **48**, 159–191.

Jain, M. K.: 1982, *Handbook of Enzyme Inhibitors*, Wiley-Interscience, New York.

Johns, E. W.: 1977, 'The Isolation and Purification of Histones', *Methods in Cell Biology* **XVI**, 183–203.

Kays, J. S., Moss, St. B., and McMaster-Kays, R.: 1979, 'Chromatographic Fractionation of Histones on Bio-Gel P', *Anal. Biochem.* **94**, 220–227.

Kleinschmidt, A. and Zahn, R. K.: 1959, 'Über Desoxyribo nucleinsäuremolekeln in Protein-Mischfilmen', *Z. Naturforschung* **14b**, 770–779.

Luck, J. M., Rasmussen, P. St., Sarake, K., and Tsvetikov, A. N.: 1958, 'Further Studies on the Fractionation of Calf Thymus Histone', *J. Biol. Chem.* **233**, 1407–1414.

Margulis, L. (ed.): 1981 *Symbiosis in Cell Evolution*, W. H. Freeman and Comp., San Francisco.

Monahan, J. J., and Hall, R. H.: 1975, 'Preparation of Chromatin from Tissue Culture Cells – A Convenient Method', *Anal. Biochem.* **65**, 187–203.

Moor, H.: 1964, 'Die Gefrier-Fixation lebender Zellen und ihre Anwendung in der Elektronenmikroskopie', *Z. Zellforsch.* **62**, 546–580.

Moor, H.: 1981, 'Die Gefrierätztechnik (2.4.2.2)', in G. Schimmel and W. Vogell (eds), Methodensammlung der Elektronenmikroskopie, Wissenschaftliche Verlagsgesellschaft, Stuttgart.

Nowotny, A.: 1979, *Basic Exercises in Immunochemistry*, Springer-Verlag, Berlin, Heidelberg, New York.

Otten, H., Plempel, M., and Siegenthaler, W. (eds): 1975, *Antibiotika-Fibel*, Georg Thieme Verlag, Stuttgart.

Pickett-Heaps, J. D.: 1974, 'The Evolution of Mitosis and the Eukaryotic Conditions', *Biosystems* **6**, 37–48.

Rizzo, P. J. and Burghardt, R. C.: 1982, 'Histone-like Protein and Chromatin Structure in Wall-less Dinoflagellate *Gymnodinium nelsonii*', *Biosystems* **15**, 27–34.

Smillie, R. M. and Scott, N. S.: 1969, 'Organelle Biosynthesis: The Chloroplast', in *Progress in Molecular and Subcellular Biology*, **1**, Springer-Verlag, Berlin, pp. 136–202.

Sonnenbichler, J. and Zetl, I.: 1975, 'The Quantitative Protein Composition of Calf Thymus Chromatin', *Hoppe-Seyerl's Z. Physiol. Chem.* **356**, 599–603.

Spring, Th. G. and Cole, R. D.: 1977, 'Chromatographic Fractionation of Histones', *Methodes in Cell Biology* **XVI**, 227–240.
van der Waerden, B. L. and Nievergelt, E.: 1956, *Tables for Comparing Two Samples by X-test and Sign Test*, Springer-Verlag, Berlin.
Wilhelm, C.: 1983, Dissertation, Fachbereich Biologie, Universität Mainz, Teil II: *Nanochlorum eucaryotum*, eine neue, marine sehr kleine Chlorophyll A und B enthaltende, einzellige Alge.
Wilhelm, C. and Wild, A.: 1982, 'Growth and Photosynthesis of *Nanochlorum eucaryotum*, a New and Extremely Small Eukaryotic Green Alga', *Z. Naturforsch.* **37c**, 115–119.
Wilhelm, C., Eisenbeis, G., Wild, A., and Zahn, R. K.: 1982, '*Nanochlorum eucaryotum*: a Very Reduced Coccoid Species of Marine Chlorophyceae', *Z. Naturforsch.* **37c**, 107–114.
Yamada, K., Murakami, H., Shirabata, S., Nishigashi, H., Shinohara, K., and Omura: 1978, 'Preparations of histone and nucleosome', *J. Fac. Agric. Kyushu Univ.* **22**, 203–210.
Zahn, R. K.: 1983, 'Measurement of the Molecular Weight Distributions in Human Muscular DNA', *Mech. Ageing Develop.* **22**, 355–379.

THE CELL CHARACTERS OF TWO HELIOFLAGELLATES RELATED TO THE CENTROHELIDIAN LINEAGE: *DIMORPHA* AND *TETRADIMORPHA*

GUY BRUGEROLLE and JEAN-PIERRE MIGNOT

Groupe de Zoologie et Protistologie, Université de Clermont, B.P. 45, 63170 Aubière, France

(Received 5 October, 1983)

Abstract. Characters of the two Helioflagellates *Dimorpha* and *Tetradimorpha* were compared to other Heliozoa particularly the Centrohelidia and to the helioflagellate *Ciliophrys*. They both have a condensed central MTOC from which originate the axopodial MTs as in the Centrohelidia. The axopodia are contractile and participate in both prey capture and slow movements of the cell. In *Dimorpha* the MTs of the axopodial cytoskeleton are arranged in 'quincunx' resulting in a quadrangular arrangement or in a squared packed arrays. This type of arrangement differs markedly from that of other centrohelidian Heliozoa where MTs are assembled in triangles and hexagons or in irregular adjacent hexagons. In *Tetradimorpha* the MTs are irregularly arranged, although they are linked together and triangular figures are visible. This latter arrangement could be considered as an incomplete and primitive system. The kinetocysts of *Dimorpha* and *Tetradimorpha* are highly structured like most of the Centrohelidia but the central body is composed of encased cylinders similar to that of *Clathrulina*. Mitochondria have contorted cristae in *Tetradimorpha* and vermicular ones in *Dimorpha* thus they differ from the plate-like cristae of the centrohelidians. They do not have any mastigonemes on their flagella.

These two helioflagellates have in common with centrohelidian heliozoa: a central MTOC, a close packing arrangement of MTs in axopods and highly structured kinetocysts. They might be considered as different levels or branches in the evolution of this lineage. They are different from the actinophryidian Heliozoa and from the helioflagellate *Ciliophrys* in respect of the characters cited above. Their study does not reveal any characters common to their supposed ancestors the chrysomonad flagellates except the orthogonal pair of kinetosomes.

1. Introduction

Ultrastructural study of many genera of heliozoa have shown several features used to describe taxonomic and evolutionary relationships in this group but some of them seem to be preponderant and give most relevant information on the evolution. Thus, the separation of Heliozoa into two main lines was based on whether the axopodia originate from microtubule organizing centers (MTOCs) scattered on the outer nuclear membrane as in Actinophridia (Ockleford and Tucker, 1973; Mignot, 1979; Patterson, 1979) or from a condensed MTOC, the centroplast, in Centrohelidia (Davidson, 1975; Bardele, 1977a,b,c). Other characters such as patterns of microtubules (MTs) in the axopods are of great importance and also the structure of kinetocysts and the shape of the mitochondrial cristae. Others, such as the appearance of swarmer cells during the cycle or the presence and morphology of scales also have their importance (see Bardele, 1977c; Cachon and Balamuth, 1977).

The fine structure study of the helioflagellate *Ciliophrys* by Davidson (1975, 1982) has shown that this genus has the main characters of the actinophryidian heliozoa and could help to elucidate the evolution of this lineage. Moreover, this genus possesses

Origins of Life **13** (1984) 305-314. 0302-1688/84/0133-0305$01.50.

some characters of the flagellates which are reminiscent of the chrysomonads, such as the presence of tubular mastigonemes and mitochondrial intracristal filaments. These observations show the interest of the helioflagellates in the heliozoan evolution, since they constitute intermediate steps between the supposed flagellate ancestor and the heliozoan organism.

In the course of a description of the ultrastructure of two other helioflagellates *Dimorpha* and *Tetradimorpha* we have tried to explore the roots of the centrohelidian lineage and to provide further data in support of Davidson's hypothesis of the flagellates origin of helioflagellates and Heliozoa.

2. Materials and Methods

Acanthocystis turfacea and *Raphidiophrys elegans* were taken as reference to Centrohelidia; they were isolated from water of ponds and cultivated by adding Protists as preys to the culture medium.

Tetradimorpha radiata and *Dimorpha mutans* were collected on the surface of fresh water containing plants after storage of a few days in the laboratory.

The cells were fixed in glutaraldehyde plus osmic acid at 1 % V/V in phosphate buffer and treated for electron microscopy by classical procedures.

3. Observations

a. *Light microscopy*: the cell of *Dimorpha mutans* (Bělař, 1926) has two flagella emerging at the anterior pole, one is anteriorly directed and embedded in mucus, the other is lateral or trailing. From the cell arise radiated and contractile axopods which serve for prey capture. The axopodial cytoskeleton diverges from a centroplast placed above the cup-like nucleus and near the pair of basal bodies at the anterior end of the cell. When disturbed, the cell withdraws its axopods, and is able to swim erratically (light and complete electron microscopy studies will be published elsewhere).

The cell of *Tetradimorpha radiata* (Brugerolle and Mignot, in press) has four anterior flagella intercalated between four axopods. The centroplast at the base of kinetosomes, gives rise to the four anterior axopods and to eight posteriorly divergent axopods. The pear-shaped cell can become round and moves randomly with its flagella.

b. *The ultrastructural characters of the helioflagellate Dimorpha and Tetradimorpha compared to the centrohelidian Heliozoa:* We shall examine the main characters used to study the relationships in Heliozoa i.e. the MTOC, the microtubular arrangement of axopodial axonemes, the morphology and structure of the kinetocysts and of the mitochondrial cristae.

The central MTOC

All the Centrohelidia have a dense and spherical centroplast situated near the center of the cell or near the pole of the conical nucleus. The center of the centroplast is generally occupied by a central disc denser than the surrounding material (Figures 1a, 2a) although it is lacking and replaced by a microfibrillar dense sphere in some centrohelidian heliozoa such as *Hedraiophrys*, *Gymnosphaera* or *Actinocoryne* (see Febvre-Chevalier, 1973a, b, 1975, 1980; Jones, 1976). When it is present such as in *Rhaphidiophrys* (Figure 1a), the central disc generally displays a tripartite structure composed of a middle plate sandwitched between two hemispheres made of a dense granular material (Bardele, 1975, 1977a, b; Davidson, 1975; Rieder, 1979). At its periphery granular material forms a shell from which arise the cytoskeletal MTs of the axopods (Figures1a, 2a). Thus the centroplast of Centrohelidia acts as a condensed and persistent MTOC similar to the granular material dispersed on the outer nuclear envelope in the Actinophryidia heliozoa (Jones and Tucker, 1981).

In *Dimorpha* the central MTOC is a microfibrillar sphere without the central disc of certain Centrohelidia (Figure 3a). In *Tetradimorpha* the central MTOC has a lens shape (Figure 4a); it is composed of an amorphous substance which seems different from the periodic structure of the dense plate which composes the central disc in *Heterophrys* (Bardele, 1977a, b).

These two helioflagellates possess a central MTOC which has structural particularities: mainly the absence of dense plate and its position near and above the nucleus at the base of the kinetosomes; like that of centrohelidians, it plays the role of a MTOC for axopodial MTs. Considering these characters the central MTOC of this two helioflagellates seems homologous to the centroplast of the Centrohelidia thus the two helioflagellates *Dimorpha* and *Tetradimorpha* can be incorporated into the centrohelidian lineage as it has been predicted by Davidson (1975). They are strictly different from the Actinophryidia (Jones and Tucker, 1982) and from the helioflagellate *Ciliophrys* (Davidson, 1982) where the axopodial MTs originate from dispersed MTOCs on the outer nuclear envelope.

The microtubular arrangement of axopodial axonemes

The second character which is common to centrohelidians Heliozoa and centrohelidian helioflagellates after the centroplast is the presence of typical axopods which represent a system for prey capture and also help the cell to move slowly or to adhere to the substratum. These axopods can contract very rapidly or slowly, probably depending on the initiation factors (Davidson, 1975; Ockleford and Tucker, 1973). It is now clearly established that axopodial shortening coincides with the depolymerization of the labile MTs which form the axopodial cytoskeleton (see Ockleford and Tucker, 1973; Edds, 1975). This is probably one of the most interesting properties shared by all the Heliozoa and also by the two helioflagellates *Dimorpha* and *Tetradimorpha*.

In the centrohelidian heliozoa *Raphidiophrys* and *Acanthocystis* microtubules of the axopodial axonemes are positioned at the apices of equilateral triangles. When two triangles are associated they form X-shaped motifs (Figure 1b). When several triangles are linked together they form irregular hexagons (Figure 1b) (Tilney, 1971; Bardele, 1977a, b). In *Heterophrys* a somewhat different microtubule assemblage leads to a similar hexagonal pattern (Bardele, 1977a, b). More recent investigations have revealed a central filament or rod in each hexagon in *Heterophrys marina* (Davidson,

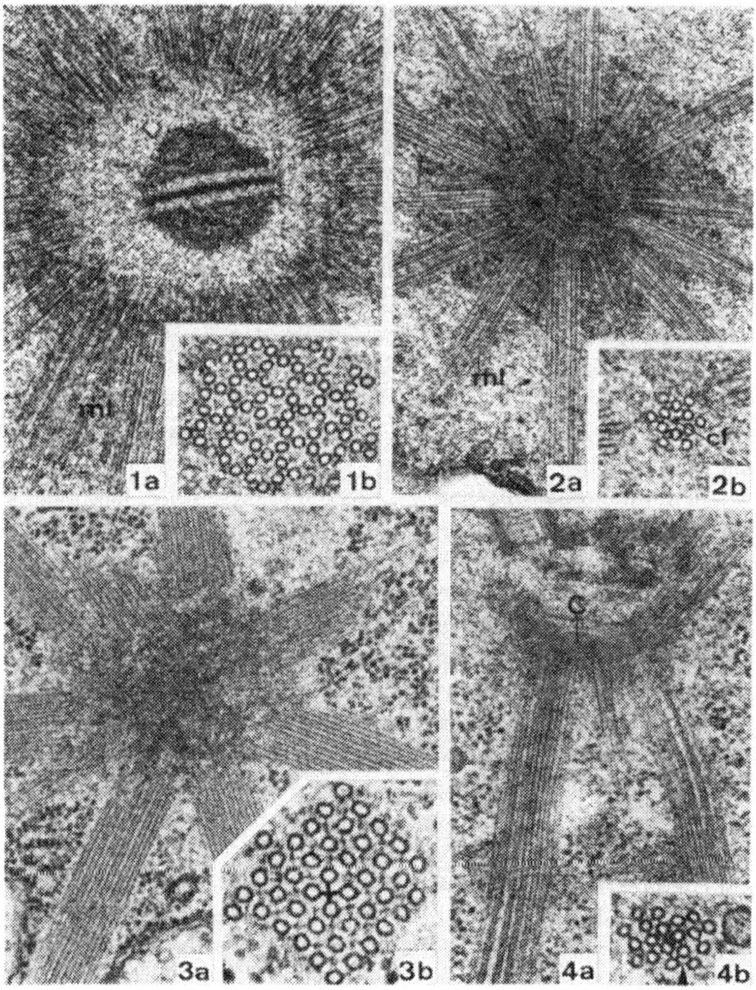

Fig. 1a. The centroplast of *Raphidiophrys elegans* comprises a tripartite central disc containing an apparent equatorial dense plate and a peripheral shell from which originate the axopodial MTs. × 60 000. b. In the axopods the MTs are arranged in triangles grouped in x-shaped figures, giving an hexagonal pattern or a packing of irregular hexagons. × 90 000.

Fig. 2a. In the centroplast of *Acanthocystis turfacea* the equatorial dense plate of the central disc is less apparent; the axopodial MTs are initiated from the dense material surrounding the disc. × 60 000. b. At the proximal end, the axopodial MTs are arranged in a single hexagon containing a central filament (cf) which is linked to the six neighbouring MTs. × 90 000.

Fig. 3a. The centroplast of *Dimorpha mutans* is composed of microfibrillar material from which arise the axopodial axonemes. × 54 000. b. Axopodial MTs are arranged in a 'quincunx' pattern that gives a quadrangular arrangement (+) or a squared packed array. × 150 000.

Fig. 4a. In *Tetradimorpha radiata* there is a lens shape centroplast (C) made of an amorphous and dense material; from its periphery arise the axopodial MTs. × 93 000. b. The MTs are irregularly arranged, some are linked by more than four bridges to the neighbouring MTs. However, triangular figures do exist (arrow). × 102 000.

1975), and in *Raphidiophrys ambiguum* (Rieder, 1979, 1980) (Figure 2b). This arrangement of MTs is maintained by four linkers positioned at 60 or 120 degrees from each other around the microtubule wall, giving a dense microtubule packing. In other centrohelidians such as *Hedraiophrys, Gymnosphaera* and *Actinocoryne* (Febvre-Chevalier, 1973a, 1975, 1980; Jones, 1976) the microtubule linkage gives rise to assymetric hexagons in which the MTs bear only 3 linkers. These two types of linkage lead to a close microtubule packing in all the centrohelidian Heliozoa which is very different from that of actinophryidian Heliozoa (Roth *et al.*, 1970; Cachon *et al.*, 1973).

In *Dimorpha* the microtubular arrangement of the axopodial axonemes is very different from that of the centrohelidian heliozoa the MTs are organized in staggered rows in a 'quincunx' pattern (Figure 3b). Within a row each MT is linked by intrarow bridges. The MTs of every other row are also linked by interrow bridges. Interrow and intrarow bridges criss-cross giving a cruciform structure. Unlike the preceding system of the centrohelidian heliozoa there are four linkers positioned at 90 degrees from each other around the microtubular wall resulting in a quadrangular pattern. This very precise arrangement resembles a paracrystalline structure, it may be described differently as parallel rows of MTs linked by the diagonal bridges in squares giving a 'square-packed arrays'. There is no visible central filaments in squares.

This pattern of microtubules is different from the possible patterns envisaged by Tilney (1971) and gives a very dense packing arrangement. The differences in the microtubular linkage between Centrohelidia and *Dimorpha* could mean that the linkers and their attachement on the subunits of the microtubular wall are different.

In *Tetradimorpha* axopodial MTs never form regular figures as in Centrohelidia or in *Dimorpha*. They are irregularly arranged and they form neither hexagons nor alternate rows with a cruciform system of linkers. Nevertheless in most cases each MT is attached to the neighbouring MTs by one to six linkers (Figure 4b). We also observed triangular figures as in Centrohelidian and the angle between two linkers is 60 degrees. No hexagonal figures were seen containing a central filament or a central MT, which are considered in the Davidson's (1975) hypotheses to be the most primitive arrangement of the axopodial MTs in the centrohelidian lineage. Probably the hexagonal pattern has originated from a more random packing such as is found in *Clathrulina* (Bardele, 1972) or in *Tetradimorpha*.

The arrangement of MTs is not homogenous in organisms that possess in common a centroplast and form the centrohelidian lineage. However this organization is very different from that of the actinophryidian lineage where MTs form two interlocking spirals (Roth *et al.*, 1970). In the actinophryidian heliozoa each Mt bears two types of thin linkers while in centrohelidia MTs have thick linkers, leading to a more tightly packed arrangement of Mts in the axopodial axoneme.

In the centrohelidian lineage a microtubular arrangement like that of *Tetradimorpha* might be considered as an incomplete and primitive system.

Kinetocyst morphology

Early studies showed that Actinophryidia have only muciferous bodies under the

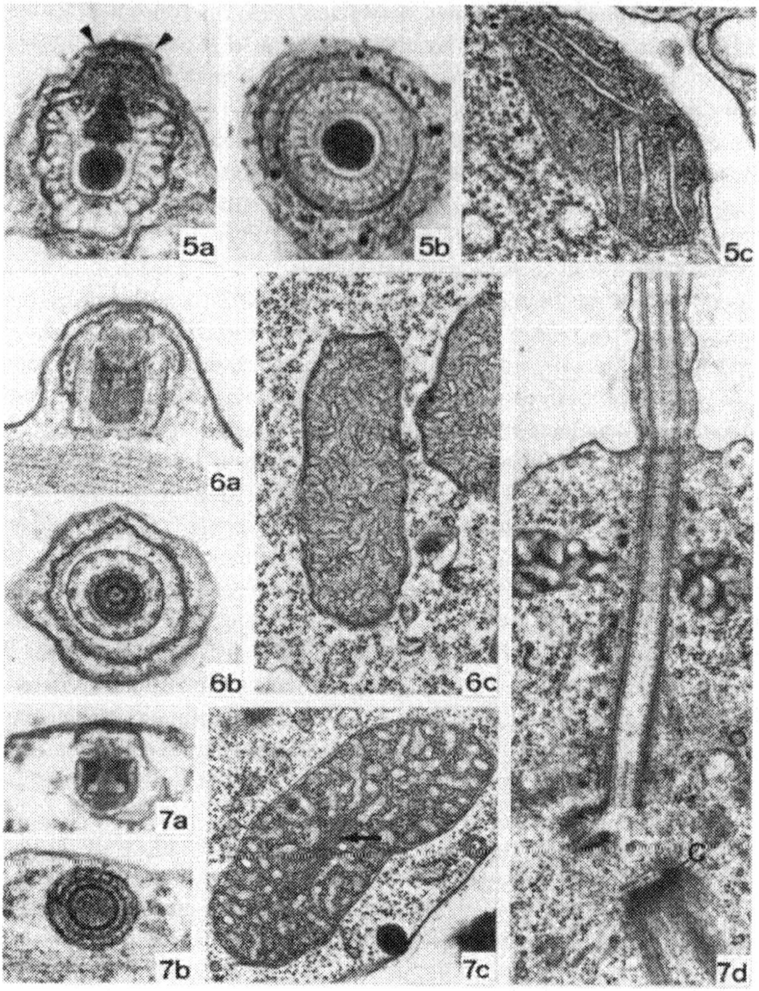

Figs. 5a, b. Longitudinal and transverse sections of a kinetocyst of *Raphidiophrys* show: the dense central body pointed outwards, the radially arranged surrounding material and the attachement zone between the kinetocyst vesicle and the cell membrane in the cone region (arrows). × 90 000. c. Mitochondrion of *Acanthocystis* has plate-like cristae disposed alongside the longitudinal axis. × 50 000.

Figs. 6a, b. The kinetocysts of *Dimorpha mutans* present a central body composed of encased cylindric tubes when seen in both longitudinal and transverse sections. × 92 000 and 90 000. c. Mitochondria of *Dimorpha mutans* show vermicular cristae. × 30 000.

Figs. 7a, b. Kinetocysts of *Tetradimorpha* are smaller than the preceding ones they display encased concentric elements pointed toward the surface. × 84 000 and 90 000. c. Mitochondrion of *Tetradimorpha* has contorted tubular cristae and the matrix contains a paracrystalline body. × 32 000. d. In *Tetradimorpha* the kinetosome of one flagella arises above the centroplast (C) it is very long and shows a conventional structure; there is no transitional helix nor tubular mastigonemes on the flagellar shaft. × 50 000.

membrane surface. Certain centrohelidians bear highly structured kinetocysts which form several little mounds on the surface of the axopods (Davidson, 1975, 1976) while in others: *Hedraiophrys*, *Gymnosphaera*, *Actinocoryne* the extrusomes seem not to be highly organized (Febvre-Chevalier, 1973b, 1975, 1980; Jones 1976). In several centrohelidians such as *Heterophrys*, *Raphidiophrys*, *Acanthocystis* kinetocysts are composed of a vesicle containing a dense body (Figures 5a, 5b). The kinetocyst body comprises a bulbous base supporting a conical shaped head directed toward the surface. The dense body is surrounded by concentric multilayers of low density material. In addition the head is attached to the membrane of the kinetocyst vesicle and to the cell membrane by rows of intramembrane particles arranged in rosettes when observed by freeze fracture techniques (Davidson, 1975, 1976; Bardele, 1976). A collar of particles that attaches the kinetocyst vesicle to the membrane underlines the base of the protrusion on the cell surface (Davidson, 1976).

In *Dimorpha* (Figures 6a, 6b) as in *Tetradimorpha* (Figures 7a, 7b) kinetocysts are very numerous both in the axopods and the cell body. They are smaller in *Tetradimorpha*. They share the general structure of those of Centrohelidia with a kinetocyst vesicle containing concentric multilayered material when seen in transverse section. In longitudinal section they display a median body of cylindrical shape without any distinct bulbous base or conical head. They look like very much to the kinetocysts of *Clathrulina* (Bardele, 1972, and Bardele, personal communication). Thus we note several major differences between the kinetocyst structure in centrohelidian heliozoa and in helioflagellates having a central MTOC. We confirm that kinetocysts in the centrohelidian lineage are highly structured compared to muciferous bodies of Actinophryidia.

Mitochondria

Typically the mitochondria of centrohelidia such as *Heterophrys*, *Acanthocystis*, *Raphidiophrys* have plate-like cristae generally oriented in parallel to the long axis of the mitochondrion (Figure 5c) (Bardele, 1977a) while the cristae are tubular in *Hedraiophrys*, *Gymnosphaera* or *Actinocoryne* (Febvre-Chevalier, 1973b, 1975, 1980; Jones, 1976). This morphology varies slightly according to the species and the physiological state of the cells.

In *Tetradimorpha* (Figure 7c) the mitochondria have distorted tubular cristae which are reminiscent of those of actinophryidian Heliozoa and chrysomonads. However, they do not contain intracristal filaments, always present in chrysomonad flagellates and also in *Ciliophrys* and pedinellids which are suggested to be related to the actinophryidian lineage (Davidson, 1975, 1982). Frequently, the mitochondrial matrix contains a paracrystalline body (Figure 7c) also shown in *Ciliophrys*.

In *Dimorpha* mitochondria have vermicular cristae which appear more distorted than in *Tetradimorpha* and there is no sign of intracristal filaments (Figure 6c).

In conclusion, the genera of the centrohelidian lineage have in common a centroplast, a dense packing arrangement of MTs forming the axopodial cytoskeleton and highly structured kinetocysts. All these characters can vary from the two primitive centrohelidian related helioflagellates to the most evolved centrohelidian heliozoa representing different levels and ramifications in the evolution of this lineage (Figure 8).

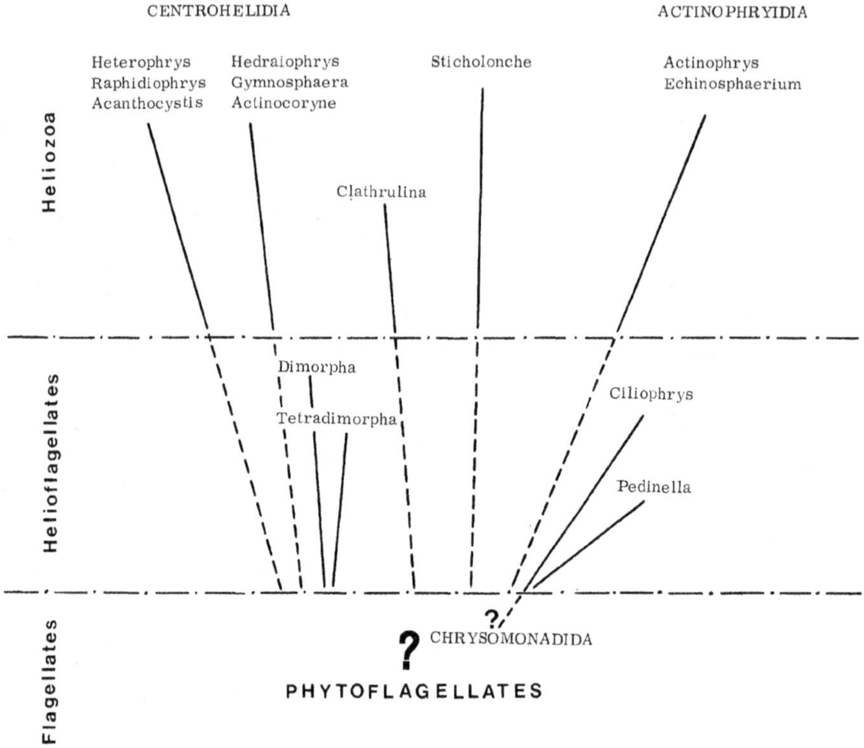

Fig. 8. Tentative evolutionary scheme showing the place of *Dimorpha* and *Tetradimorpha* among the others helioflagellates and among the Heliozoa which have probably originated from the Phytoflagellates.

c. *Comparison between centrohelidian related helioflagellates and the supposed Flagellate ancestor:* The study of *Ciliophrys* by Davidson (1975, 1982) has revealed some characters also shared by chrysomonads i.e. tubular mastigonemes on flagella and an intracristal filament newly observed in mitochondria. It may be asserted that the two centrohelidian related helioflagellates studied do not retain any chrysomonad character except for the orthogonal pair of kinetosomes present in the heterokont group of Flagellates. On the flagellar surface we can only detect a coat but no tubular mastigonemes (Figure 7d). Moreover the transitional helix present in chrysomonads, which is a structure highly preserved in the evolution (Hibberd, 1979), does not exist in centrohelidian related helioflagellates or in pedinellids.

It might have been expected that the flagellar cytoskeleton of the helioflagellate could have retained some fibers typical of the flagellate ancestor. In fact, there is no trace of a rhizoplast associated to the golgi complex and to the nuclear envelope like in chrysomonads. In centrohelidian related helioflagellates there are many Golgi bodies radially dispersed which are apparently independent from the flagellar roots.

Although centrohelidian heliozoa are able to build siliceous scales or spicules like several chrysomonads, the two centrohelidian related helioflagellates do not have any. Division figures of *Dimorpha* show that the nuclear envelope disappears and spindle MTs are polarized by a large zone around the pairs of kinetosomes while in chrysomonads there is a fiber attached to the kinetosomes which polarizes the spindle (Slankis and Gibbs, 1972).

Though some chrysomonad cells have axopod-like organelles they never contain bundles of microtubules but rather microfilaments. They are termed filopodia in amoeboid chrysomonad such as *Rhizochromulina* (Hibberd *et al.*, 1979) or filose pseudopodia in Amoeba such as *Nuclearia* (Mignot and Savoie, 1979) or *Nucleosphaerium* (Cann and Page, 1979). However, more recent investigations on the Foraminifer *Allogromia* have revealed a microtubular and microfibrilar cytoskeleton in the reticulopodial network (Travis and Allen, 1981) but the microtubular pattern is not similar to one of those present in Heliozoa (see Bardele, 1977d).

These two centrohelidian related helioflagellates appear already engaged toward the centrohelidian evolutionary lineage and they might have lost many characters of the flagellate ancestor (Figure 8). As the helioflagellates represent intermediate steps in the evolution of a flagellate state to a heliozoan state they may help us to discover the origin of the different lineages of Heliozoa. This origin is probably polyphyletic and supposed to be among the Phytoflagellates (Figure 8).

References

Bardele, C. F.: 1972, Cell Cycle, Morphogenesis and Ultrastructure in the Pseudoheliozoan *Clathrulina elegans*, *Z. Zellforsch.* **130**, 219–242.

Bardele, C. F.: 1975, The Fine Structure of the Centrohelidan Heliozoan *Heterophrys marina*, *Cell Tiss. Res.*, **161**, 85–102.

Bardele, C. F.: 1976, Particle Movement in Heliozoan Axopods Associated with Lateral Displacement of Highly Ordered Membrane Domains, *Z. Naturforsch.* **31c**, 190–194.

Bardele, C. F.: 1977a, Comparative Study of Axopodial Microtubule Pattern Control in the Centrohelidian Heliozoa *Acanthocystis, Raphidiophrys and Heterophrys*, *J. Cell Sci.* **25**, 205–232.

Bardele, C. F.: 1977b, Organization and Control of Microtubule Pattern in Centrohelidian Heliozoa, *J. Protozool.* **24** (1), 9–14.

Bardele, C. F.: 1977c, Evaluation of Ultrastructural Features in the Classification of Heliozoan Actinopods. *Abst. 75*, in S. H. Hutner (ed.), *Proceedings of the Vth Int. Cong. Protozool, New York*, Pace University, New York.

Bardele, C. F.: 1977d, On the Involvement of the Plasma Membrane in Saltatory Motion of Particles in the Rhizopoda. *Abst. 54*, in S. H. Hutner (ed.), *Proceedings of the Vth Int. Cong. Protozool., New York*, Pace University, New York.

Bělař, K.: 1926, *Der Formwechsel des Protistenkerne*, G. Fischer, Jena, ed.

Brugerolle, G. and Mignot, J. P.: 1983, Caractéristiques Ultrastructurales de l'Hélioflagellé *Tetradimorpha* (Hsiung) et leur Intérêt pour l'Etude Phylétique des Héliozoaires, *J. Protozool.* (in press).

Cachon, M. and Balamuth, W.: 1977, The Sarcodina, 190–196, in S. H. Hutner (ed.), *Proceedings of the Vth Int. Congress of Protozoology, New York*, Pace University, New York.

Cachon, J., Cachon, M., Febvre-Chevalier, C., and Febvre, J.: 1973, Déterminisme de l'Edification des Systèmes Microtubulaires Stéréoplasmiques d'Actinopodes, *Arch. Protistenk.* **115**, 137–153.

Cann, J. P. and Page, F. C.: 1979, *Nucleosphaerium tuckeri* nov. gen, nov. sp. A new freshwater filose Amoeba without Motile Form in a new Family *Nucleariidae* (*Filosea: Aconchulina*) Feeding by Ingestion only, *Arch. Protistenk.* **122**, 226–240.
Davidson, L. A.: 1975, *Studies of the Actinopods Heterophrys marina and Ciliophrys marina: Energetic and Structural Analysis of their Contractile axopodia, General Ultrastructure, and Phylogenetic Relationships*, Ph. D. thesis, University of California, Berkeley.
Davidson, L. A.: 1976, Ultrastructure of the Membrane Attachment Sites of the Extrusomes of *Ciliophrys marina* and *Heterophrys marina* (Actinopoda), *Cell Tissue Res.* **170**, 353–365.
Davidson, L. A.: 1982, Ultrastructure, Behavior, and Algal Flagellate Affinities of the Helioflagellate *Ciliophrys marina*, and the Classification of the Helioflagellates (Protista, Actinopoda, Heliozoa), *J. Protozool.* **29** (1), 19–29.
Edds, K. T.: 1975, Motility in *Echinosphaerium nucleofilum*. I. An analysis of Particle Motions in the Axopodia and a Direct Test of the Involvement of the Axoneme, *J. Cell Biol.* **66**, 145–155.
Febvre-Chevalier, C.: 1973a, Un Nouveau Type d'Association des Microtubules Axopodiaux chez les Héliozoaires, *Protistologica*, **9** (1), 35–43.
Febvre-Chevalier, C.: 1973b, *Hedraiophrys hovassei*, Morphologie, Biologie et Cytologie, *Protistologica* **9**, 503–520.
Febvre-Chevalier, C.: 1975, Etude Cytologique de *Gymnosphaera albida* Sassaki 1894 (Héliozoaire Centrohelidia), *Protistologica* **11**, 331–344.
Febvre-Chevalier, C.: 1980, Behaviour and Cytology of *Actinocoryne contractilis* nov. gen. nov. sp. A new stalked Heliozoa (Centrohelidia), *J. Mar. Biol. Assoc. U.K.* **60**, 909–928.
Hibberd, D. J.: 1979, The Structure and Phylogenetic Significance of the Flagellar Transition Region in the Chlorophyll c-Containing Algae, *Biosystems* **11**, 243–261.
Hibberd, D. J. and Chretiennot-Dinet, M. J.: 1979, The Ultrastructure and Taxonomy of *Rhizochromulina marina* gen. et sp. nov., and amoeboid marine Chrysophyte, *J. Mar. biol. U.K.* **59**, 193–197.
Jones, J. C. R. and Tucker J. B.: 1981, Microtubule-organizing Centres and Assembly of the Double-Spiral Microtubule Pattern in Certain Heliozoan Axonemes, *J. Cell Sci.* **50**, 259–280.
Jones, W. C.: 1976, The ultrastructure of *Gymnospaera albida* Sassaki, a marine axopodiate protozoon, *Phil. Trans. Soc. London* **275**, 349–384.
Mignot, J. P.: 1979, Etude Ultrastructurale de la Pédogamie chez *Actinophrys sol* (Héliozoaire). I. La division progamique, *Protistologica* **15**, 387–406.
Mignot, J. P. and Savoie, A.: 1979, Observations Ultrastructurales sur *Nuclearia simplex* Cienkowski (Protozoa, Rhizopodea, Filosia), *Protistologica* **15**, (1), 23–32.
Ockleford, C. D. and Tucker, J. B.: 1973, Growth, Breakdown, Repair, and Rapid Contraction of Microtubular Axopodia in the Heliozoan *Actinophrys sol.*, *J. Ultrastruct. Res.* **14**, 369–387.
Patterson, D. J.: 1979, On the Organization and Classification of the Protozoon, *Actinophrys sol* Ehrenberg 1830, *Microbios* **26**, 169–208.
Rieder, C. L.: 1979, High-voltage Electron Microscopy Reveals New Components in the Axonemes and Centroplast of the Centrohelidian *Raphidiophrys ambigua*, *J. Cell Sci.* **40**, 215–234.
Rieder, C. L.: 1980, Structure of the Mammalian Kinetochore and the Centrohelidian Centroplast as Revealed by Bernhard's Staining of Thick Sections. In M. de Brabander and I. de Mey (eds.), *Microtubules and Microtubules Inhibitor*, Elsevier Publ. Amsterdam, pp. 311–324.
Roth, L. E., Pihlaja, D. J., and Shigenaka, Y.: 1970, Microtubules in the Heliozoan Axopodium. I. The gradion Hypothesis of Allosterism in Structural Proteins, *J. Ultrastruct. Res.* **30**, 7–37.
Slankis, T. and Gibbs, S. P.: 1972, The Fine Structure of Mitosis and Cell Division in the Chrysophycean Alga *Ochromonas danica*, *J. Phycol.* **8** (3), 243–256.
Tilney, L. G.: 1971, How Microtubule Pattern Are Generated. The Relative Importance of Nucleation and Bridging of Microtubules in the Formation of the Axoneme of *Raphidiophrys*, *J. Cell Biol.* **51**, 854–857.
Travis, J. L. and Allen, R. D.: 1981, Studies on the Motility of the Foraminifera. I. Ultrastructure of the Reticulopodial Network of *Allogromia laticollaris* (Arnold), *J. Cell Biol.* **90**, 211–221.

AXONEMAL MICROTUBULE PATTERN OF *CIENKOWSKYA MERESCHKOVSKYI* AND A REVISION OF HELIOZOAN TAXONOMY

COLETTE FEBVRE-CHEVALIER and JEAN FEBVRE

Laboratoire de Protistologie Marine, Station marine, 06230 Villefranche sur Mer, Franche; Université de Nice, UER, FST, Parc Valrose, 06034 Nice-Cedex, France

(Received 5 October, 1983)

Summary. *Cienkowskya mereschkovskyi* is a spherical heliozoan with a mucous stalk and a cell coat with spicules randomly distributed.

The axonemal microtubules arise from a central MTOC with a centroplast. The microtubular pattern displays a triradial symmetry and is characterized in cross-section by one central hexagon with alternate angles of 106° and 134°. From this slightly irregular central pattern, convex branches made of a series of microtubules and linkers diverge. Any three microtubules (Mt's) along the branch form 166° angles. Around the central hexagon links between successive branches always delineate 3 pairs of enneagons and 3 heptagons, while the peripheral polygons are less strictly defined. None of the previously published theories on Mt-pattern formation are sufficient to totally explain such a complex pattern.

A comparison between the fine structure of *Cienkowskya* and other Heliozoa with reference to very recent data leads us to try to amend the existing terminology for the substructure of the MTOC. An attempt at a more coherent classification is also made using electron microscopy criteria with new arguments compared to existing ones (Febvre-Chevalier, 1982, 1983). A taxonomical key comprising several new names is proposed.

1. Introduction

Ultrastructural and physiological studies have emphasized the pre-eminent role of microtubules (Mt's) in the maintenance of cell shape, morphogenesis and motility processes (Dustin, 1978). In Radiolaria, Acantharia and Heliozoa (Protozoa, Actinopoda) Mt-patterns exhibit the greatest complexity and diversity (Cachon and Cachon, 1972a, b, 1974; Cachon *et al.*, 1973; Febvre-Chevalier, 1973a; Bardele, 1977a). It now appears that Microtubule Organizing Centres (MTOC's, Pickett-Heaps, 1969) and Mt-patterns are valuable characters for evolutionary taxonomy (Febvre-Chevalier, 1982). This paper is concerned with the cytological study of the heliozoan *Cienkowskya mereshkovskyi* (Cienk.), with particular attention to its Mt-organization, its MTOC and its taxonomic relationships with other species previously described (Tilney, 1971; Davidson, 1975; Bardele, 1977a, b).

The marine centrohelidan heliozoan *Cienkowskya mereshkovskyi* (Cienkovsky, 1881) syn. *Wagneria mereshkovskyi* Cienkovsky, 1881 was first described by Cienkovsky from the White Sea and later by Villeneuve (1937) (from brackish water in the south of France) who described its morphology and binary fission.

Origins of Life **13** (1984) 315-338. 0302-1688/84/0133-0315$03.60.

2. Material and Methods

Samples were collected from shallow water in Villefranche-sur-Mer (France). Organisms were found attached by a mucous stalk to hard substrates or floating just over the benthos. Cultures were grown in polystyrene petri-dishes containing filtered sea water (0.22 µm) and kept at 14 °C in a culture chamber. They were maintained in culture for some weeks and fed with *Prasinocladus* and *Dunaliella*. After 4 to 5 binary fissions the cells decreased in size and death occured after a few days.

Fixation for electron microscopy (EM) was made using the Karnovsky (1965) method modified by adjusting the osmotic pressure to that of the sea water with phosphate buffer. The cells were kept for one hour in this solution, then rinsed and post-fixed for one hour with 2% OsO_4 in phosphate buffer. After washing, cells were dehydrated in alcohol and acetone or propylene oxide and embedded in Spurr resin (Spurr, 1969). Thin sections were made with a Reichert ultramicrotome, stained with uranyle acetate and lead citrate, and electron micrographs were taken in the HU 12 microscope of the University of Nice (France). Chemicals were provided by Sigma, St. Louis, Missouri, U.S.A.

3. Results

3.1. Light microscopical observations

Cienkowskya is a spherical cell 40 to 50 µm in diameter from which radiate a few hundred axopods. The cell body is enclosed in a capsule which is at the extremity of a long rigid stalk (Figure 1). It reproduces by binary fission; no multiple fission or sexual reproduction have been observed.

The Cytoplasm

The cytoplasm is not differentiated into endoplasm and ectoplasm and the cell membrane is not deeply invaginated as reported in other heliozoa (Febvre-Chevalier, 1973b). Two regions can be distinguished within the cytoplasm, a clear central region which contains one large spherical body and a peripheral one. The central region is made of three parts: a central granule which stains strongly with iron hematoxylin or eosin, a radially striated sphere called centroplast sensu Dobell (1917) which surrounds the central granule and a hetero geneous cytoplasm which contains the preceeding structures (Figure 2). The peripheral region contains a large number of food vacuoles and a large nucleus 12 to 15 µm in diameter.

Axopods

Undisturbed cells possess long thin granule-studded axopods which are immobile except when disturbed or during secretory phases when refringent granules can be seen moving from the edge of the cell body along the axis of the axopods. If touched by prey or a glass needle, they contract quickly although contraction-velocity is weaker than in

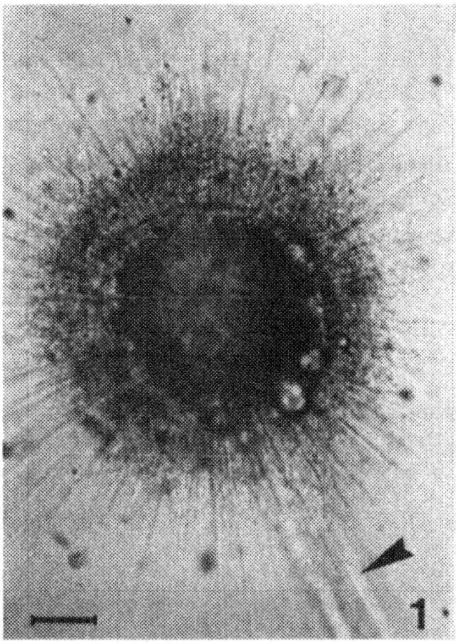

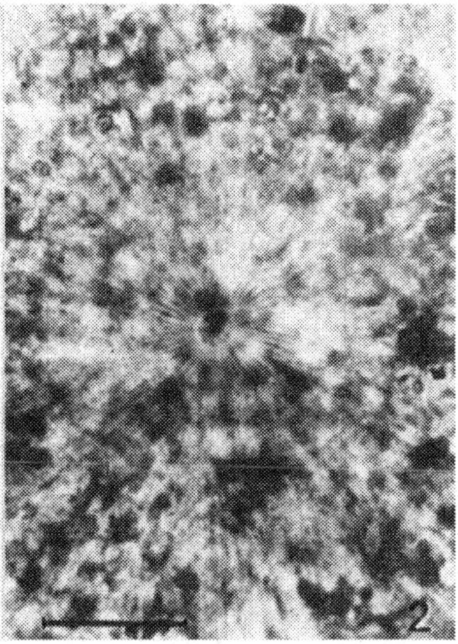

Fig. 1. Light micrograph of *Cienkovskya mereschkovskyi* with its stalk (arrow-head). Bar = 10 μm.

Fig. 2. Light micrograph of the central region and surrounding cytoplasm. Bar = 10 μm.

other species in which it can reach 200 length s^{-1} (Davidson, 1975; Febvre-Chevalier, 1981; Febvre-Chevalier and Febvre, 1980).

Capsule and Stalk

The capsule surrounding the cell (Figure 1) consists of a mucous cell coat and numerous long spicules which are probably chitin-like and not siliceous as reported earlier (Villeneuve, 1937). Their arrangement is at random rather than radial. The capsule is located at the top of a long empty peduncle (1 to 5 mm) which is attached to the substrate. Detachment from the substrate with tweezers does not induce contraction of the cell body. Thus this structure is completely different from the cytoplasmic stalk of *Actinocoryne contractilis* (Febvre-Chevalier, 1980; 1981) and *Wagnerella borealis* (Zuelzer, 1909). When disturbed the cell leaves its shell and floats for one hour or more while secretion of a new capsule occurs. When a suitable substrate is found, active motion of secretory granules is observed along the axial rods of the axopods oriented towards the substrate and in less than 12 hours a new stalk is built. Building of the capsule is performed by active secretion and transport of mucus and spicules. At the beginning of its formation the capsule lies close to the cell surface and as soon as it becomes thicker it is driven off by the axopods so that the cell is separated from its shell by a space measuring 10 to 20 μm.

3.2. Electron microscopical observations

Low magnification electron micrographs (Figures 3, 4, 5) show the fine organization of *C. mereshkovskyi*. The cytoplasm is organized in concentric layers around the central granule: successively, the dense granule, a clear zone free from organelles, an electron translucent plasma including the bases of the axopodial axonemes, a large zone of cytoplasm containing most of the organelles (Figure 3) and the nucleus (Figure 4) and finally the peripheral part of the cell filled with food vacuoles (Figure 5). Outside the cell, extrusomes can be seen which are scattered in the space between cell and mucous coat (Figure 5).

Golgi System

The golgi system consists of a large number of dictyosomes located in the subcentral cytoplasm (Figure 6). Seven to nine saccules are regularly seen which emit laterally

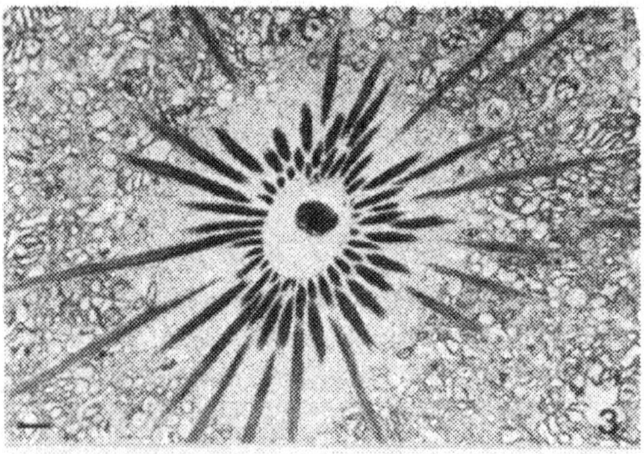

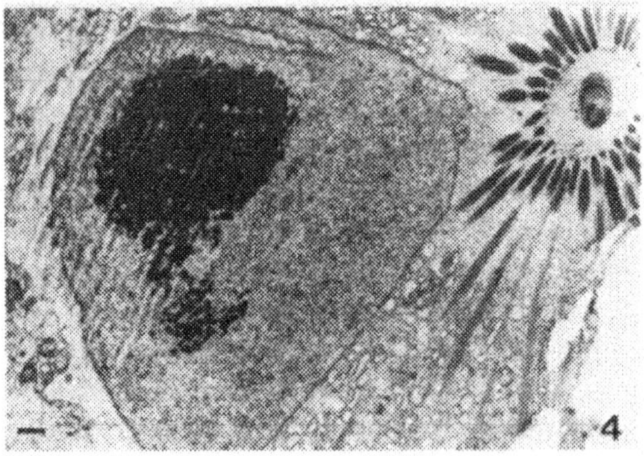

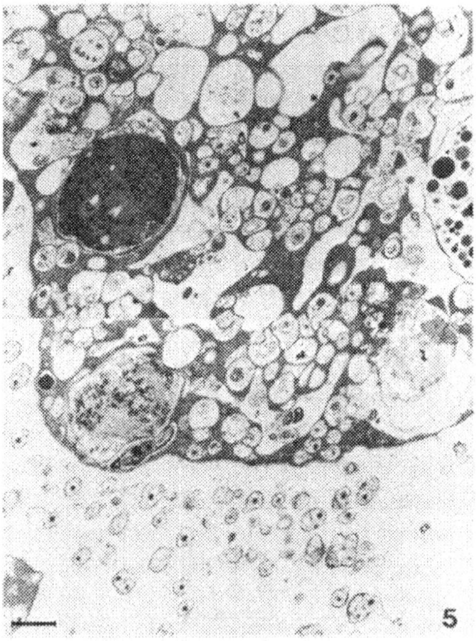

Fig. 3. Survey electron micrograph of the central region showing the centroplast, the centrosphere with the exclusion zone, the interaxonemal substance (IAS) and the axonemes. Bar = 1 μm.

Fig. 4. MTOC and nucleus just after the cell division. Bar = 1 μm.

Fig. 5. Micrograph of the cytoplasm in the peripheral region of the cell. Note the food vacuoles at different stages of the digestion and the lacunar aspect of the cytoplasm. Outside the cell numerous kinetocysts are visible. Bar = 1 μm.

small vesicles containing a fuzzy material. Large vesicles containing spherical granules are generated at the upper face of the dictyosomes. The saccules include very dense ring-shaped deposits.

Extrusomes

Two kinds of extrusomes have been shown in *Cienkowskya*, mucocysts involved in secretion of cell coat and stalk, and kinetocysts implicated in food capture. As previously described in other species, extrusomes result from the golgi system. Kinetocysts have a complex structure and display a progressive maturation during migration from the subcentral cytoplasm to the superficial region (Figures 7 and 8). When extrusion occurs the membranes of both the kinetocysts and the cell collapse and

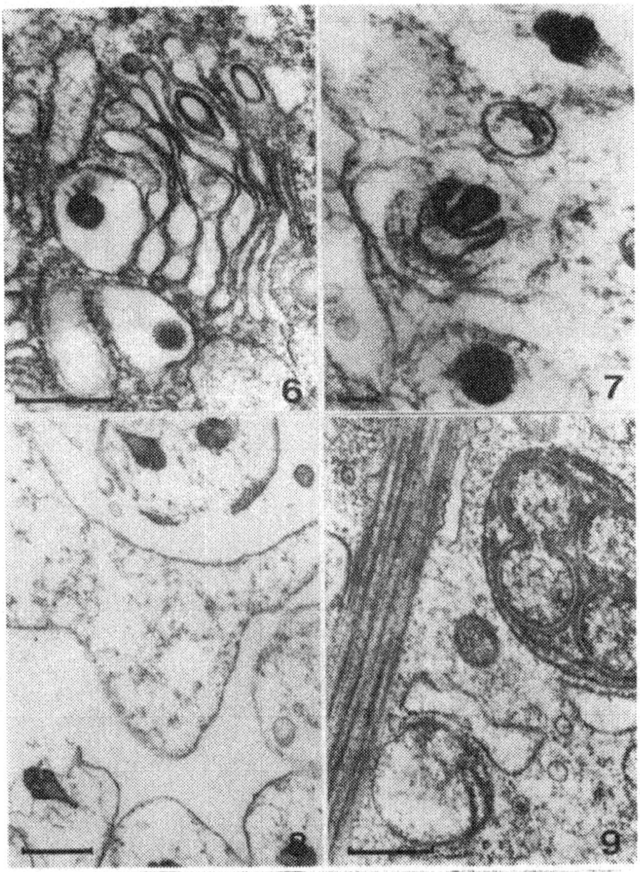

Fig. 6. Cross-section through dictyosomes displaying tiny vesicles, and large vesicles, one of them containing one dense granule. Bar = 0.5 μm.

Fig. 7. Prekinetocysts at three stages of their formation. Bar = 0.1 μm.

Fig. 8. Mature kinetocysts. Bar = 0.5 μm.

Fig. 9. Mitochondrion invaded by dividing bacteria. Lamellar cristae are closely fitted to the inner membrane. Bar = 0.5 μm.

toxic substances are ejected as a thin filament. Such generation and extrusion of kinetocysts have already been described in the heliozoa by Bardele (1975) and Davidson (1975), and freeze fracture data have shown an 8-particle rosette in the region of attachment of the organelle to the cell membrane (Davidson, 1976).

Mitochondria

Mitochondria are large organelles with lamellar cristae. Sometimes bacteria invade the mitochondrial matrix, where they divide and induce rupture of the organelle (Figure 9). Such a desease has been previously observed (unpublished data) and could explain why it is so difficult to maintain heliozoa in culture more than three weeks.

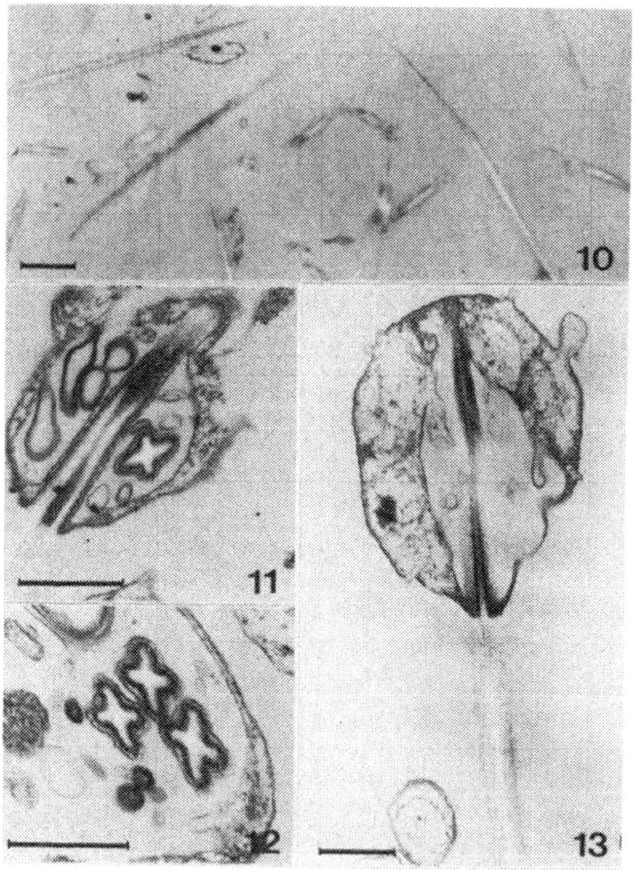

Fig. 10. External electron-lucent spicules showing the proteinacious matrix Bar = 1 µm.
Fig. 11. Spicule generating organelle (SGO). Two spicules are cut at right angle. Bar = 0.5 µm.
Fig. 12. Cross section through three spicules. Bar 0.5 µm.
Fig. 13. Spicule generating organelle (SGO) and the proteinacious matrix of the spicule. Bar = 1 µm.

Spicules

The mucous capsule of *Cienkowskya* includes a large number of spindle-shaped spicules which are dissolved immediately by sulfuric acid and are easily cut with glass knives, indicating that they are probably not siliceous as Villeneuve thought. In longitudinal and cross sections they appear to be hollow and covered with a proteinacious coat (Figures 10 and 13). In cross-sections spicules are cruciform (Figures 11 and 12). They are produced by 'spicule generating organelles' (SGO) (Figure 11) similar to those of *Heterophrys* (Bardele, 1975; Davidson, 1975). A SGO is composed of an invagination of the cell membrane surrounded by a large vacuole. The base of the spicule which is located within this enfolding, appears to be made of an electron-dense substance, whereas the distal part (i.e. outside of the cell invagination) is electron-translucent. Often several spicules are elaborated simultaneously by one SGO (Figure 11). Accumulation of ring-like substances similar to those found in the saccules of the dictyosomes are observed in the vacuole surrounding the SGO but we have no direct indication that they are used for spicule base elaboration.

Centroplast and Central Electron-Translucent Plasma

The central granule or centroplast consists of a tripartite disc surrounded by an electron-dense material 1.5 μm in diameter (Figure 20). It is included in an electron-translucent plasma called centrosphere which is itself divided into two regions: (1) an exclusion zone of 1 to 1.5 μm close to the centroplast which is free from organelles and (2) a similar plasma 3 to 3.5 μm thick in which the axonemes are rooted (Figures 14 and 15). Thin filaments connect the rods of axonemal Mt's to one another and to the centroplast so that this region resembles a network (Figure 14).

Axopodial Axonemes and Mt-Pattern

Proximal ends of the axonemal rods are more dense than distal ends. Axonemes appear to be made of closely packed Mt's embedded in an electron-translucent material. In the proximal region the number of Mt's increases progressively so that in longitudinal section the basis of the axoneme has a sharp conical shape. Central Mt's are nearer to the centroplast than peripheral ones (Figure 15).

Cross-sections through the axopodes show that each axoneme is made of about 150 Mt's which gives a mean diameter of 0.7 μm (Figure 16). Each Mt has the usual size (25 nm in diameter) and bridges between Mt's measure 7 nm. The microtubular arrangement is the most complex found until now in the heliozoa although it is built around a central hexagon similar to that reported in several species of heliozoa (Tilney, 1971; Davidson, 1972, 1975; Bardele, 1975, 1977a, b).

In the center of each axoneme, 6 Mt's linked by bridges delimit a slightly irregular hexagon whose angles measure alternately 106° and 134°. This hexagon is surrounded by 2 or a maximum of 3 adjacent series of polygons, namely equilateral triangles, irregular enneagons (9 edges and angles), heptagons and a variable number of squares and lozenges (Figures 16 and 17). Polygons result from the arrangement of Mt's and bridges according to divergent curved branches (also called arcs) whose concavity is

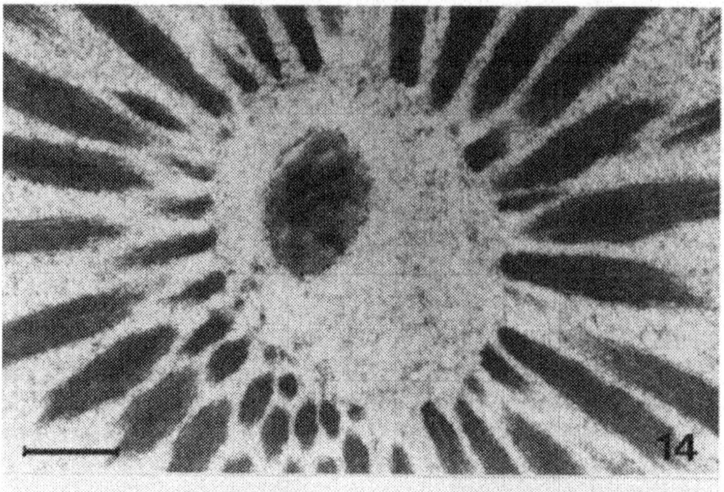

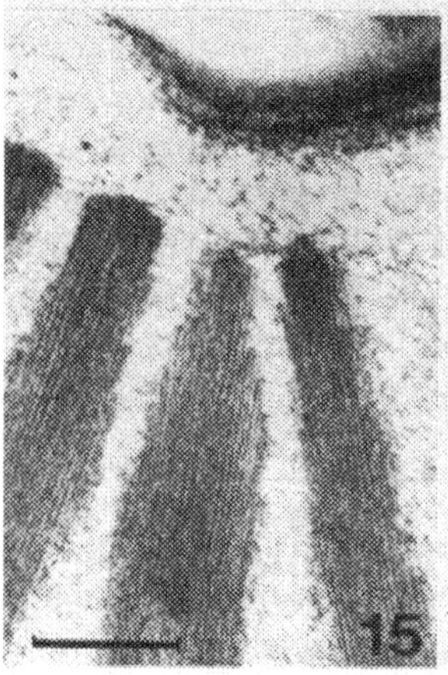

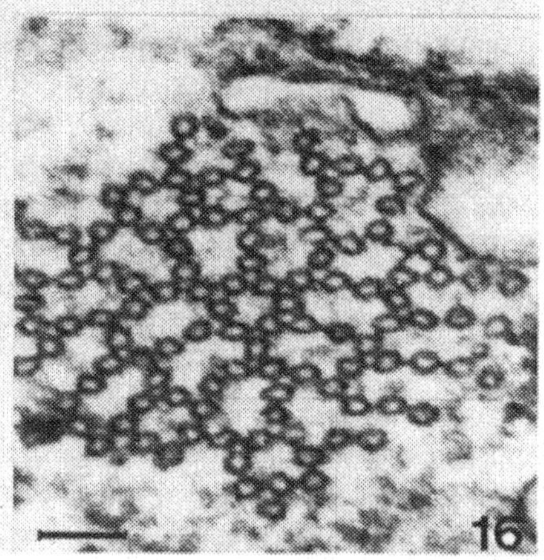

Fig. 14. Oblique section through the MTOC displaying the centroplast, the exclusion zone, dense bases of the axonemes and the inter-axonemal substance. Bar = 1 μm.

Fig. 15. Detail of the Mt. rooted in the centrosphere. Note the progressive increase in the diameter of the axoneme by addition of Mts. Bar = 0.5 μm.

Fig. 16. Cross-section through the axonemal pattern. Bar = 0.1 μm.

always faced in the same direction. Angles between successive bridges are constant and measure 166°. This arrangement induces a tri-radial symmetry of the whole pattern as shown by the labelling of Figure 17. To understand the geometry of this highly complex pattern, we have labelled A, B, C, corners of the central hexagon which measure 106° and I, II, III, those which measure 134° (Figure 17). From A, two curved branches diverge forming an angle of 60°. Each of them is made of 6 Mt's and bridges. Mt's are named A_a, A_b, A_c, A_d, A_e, A_f, A_g and A_α, A_β, A_γ, A_δ, A_ε, A_ζ. A_a and A_α are superimposed and correspond to the same Mt A of the central hexagon. The Mt I also corresponds to A_β with which it merges. From I, only one arc is built with 6 Mt's and links, numbered

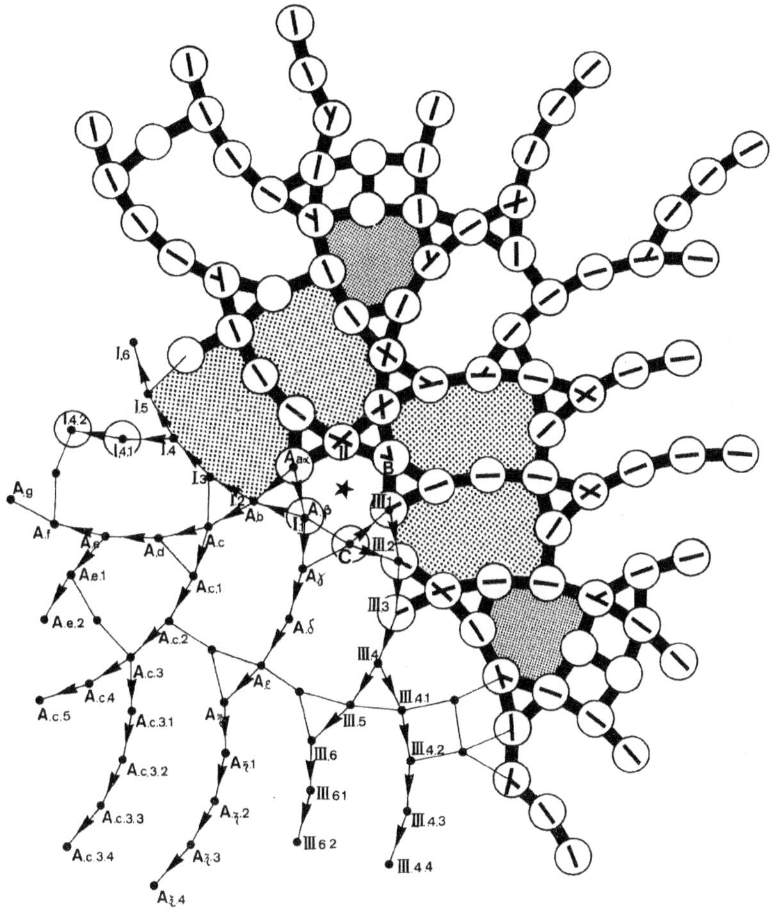

Fig. 17. Interpretative drawing of the axonemal Mt pattern in cross section. For the explanations, see in Results: axopodial axonemes and Mt-pattern.

I_1 to I_6. It makes a 60° angle with the segment $A_\alpha A_\beta$ and exhibits a 166° angle between two adjacent bridges. The branches $A_a - A_g$ and $I_1 - I_6$ cross at I_2 A_b delimiting an equilateral triangle which is adjacent to the hexagon. This situation is repeated for each pair of homologous arcs, furthermore a bridge links Mt's C to A_γ thus delineating a six-pointed star around the central hexagon. From precise locations on the convex face of each of these lettered and numbered arcs, secondary branches, similar to the primary ones (166° angle between adjacent bridges) are generated and form an angle of 60° with the primary arcs ($A_c - A_{c.5}$; $A_\zeta - A_{\zeta.4}$; $III_4 - III_{4.4}$ for example). Tertiary branches having similar characteristics are grafted onto the secondary branches. They originate, for instance, from $A_{c.3}$ and $I_{4.4}$, and consist of 3, 4 or 5 Mt's.

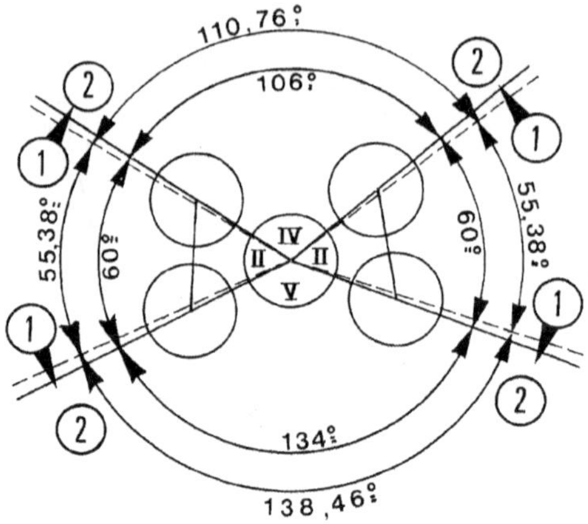

Fig. 18. Geometric explanation of the angular compensation at the level of the central hexagon angles and at that of triangles forming the 6-pointed star. 1 and 2 indicates the hypothetical displacement of linkers.

Links whose concavity is turned towards the central hexagon and which are made of one Mt and 2 bridges at 166° angles connect adjacent primary and secondary arcs at precise levels ($A_{c.2}-A_{\varepsilon}$; $A_{\varepsilon}-III_5$). They delineate around the six-pointed star-hexagon a stable geometry i.e. 3 pairs of equal enneagons. Similar links connect $A_{e.1}$ to $A_{c.3}$; bridges also link $A_d-A_{c.1}$ so that 3 heptagons are always observed. From this level, and in particular, at the periphery of the axoneme, the shape of the polygons varies due to variation in the position of links (made of 1 Mt and 2 bridges at 166° angles). For instance $III_{4.1}$ linked to C_e delineates one heptagon instead of one enneagon. Sometimes hendecagons (11 edges and angles) are built. Successive links can also be connected by one bridge and thus delineate squares and lozenges. In most cases, there is no link between adjacent arcs at the periphery of the pattern resulting in open bays. In some cases accidental modification occur in linkage, for example in $A_{e.1}-A_{e.2}$ (Figure 17) the curvature is turned counterclockwise whereas all the arcs are turned clockwise.

4. Discussion

Fine structure of *Cienkowskya mereshkovskyi* and particularly organization of the arrays of Mt's which stiffen the axopods suggest three areas of interest (1) those related to Mt-pattern formation with reference to different hypotheses of nucleation, (2) those concerning the MTOC and (3) those concerning taxonomic relationships with the other heliozoa.

4.1 Pattern formation and nucleation

Most microtubules studied thus far are composed of 13 tubulin subunits (Dustin, 1978). Although the quality of fixation in this study does not allow us to count the

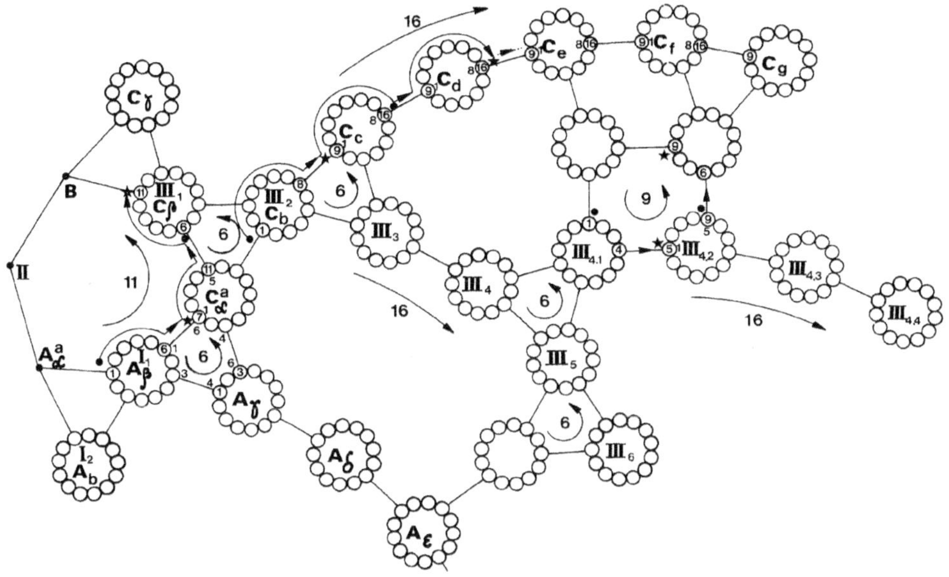

Fig. 19. Diagram to illustrate the formation of enneagons and heptagons from the theory of "linker nucleation hypothesis". Gradients are indicated with arrows.

number of Mt-subunits, angles measured on electron micrographs suggest that Mt's are also built with 13 subunits in this organism.

How is the Central Hexagon Built?

A central regularly distorted hexagon with successive angles of 134° and 106° and with equal linkers has already been described in *Raphidiophrys*, *Acanthocystis* and *Heterophrys* (Bardele, 1977a, b). In *Cienkowskya* one can also suggest that the large angles (134° observed in cross-section) include 5 subunits and the small ones (106 measured) 4 subunits (Figure 19). But theoretical angular values for 5 and 4 subunits are 138.46° and 110.76°, respectively. The sum of these 6 theoretical angles is 747.66°, or 27.69° larger than the sum of the angles of a closed hexagon ($(6-2)\pi = 4\pi = 720°$). This difference of 27.69° can include one more subunit between its edges. Therefore to have a closed hexagon each angle must be 27.69° : 6 = 4.61° smaller than the theoretical value associated with 5 and 4 subunits. In other terms, to close the central hexagon, each link must be displaced by 2.3°. Such a stressed configuration is compatible with 13 subunits each linker being slightly displaced.

How Are the Curved Branches Constructed?

Along the curved branches, the angles delineated by two successive links have exactly the same value as the theoretical ones i.e. 166°. They include 6 subunits between their linkers and 7 subunits for the complemantery angle (Figure 19). Why are the observed angles in exact agreement with the theoretical angles? At the level of the Mt's A, B, C subunits are in 4-2-5-2 ratio. The following table gives theoretical angles, observed angles, and the angular differences in each case.

Number of subunits	Theoretical angles	Observed angles	Angular difference
4	110.76°	106°	− 4.76°
2	55.38°	60°	+ 4.62°
5	138.46°	134°	− 4.46°
2	55.38°	60°	+ 4.62°

The difference between theoretical and observed angles is constant; the 4.76° negative difference between the theoretical and observed angles with 4 subunits, and a similar difference (4.46°) in angle with 5 subunits are compensated by an increase of 9.24° in the two theoretical angles with 2 subunits. This allows us to see that whatever the method of measurement, one has a 166° angle. $A_{\beta}C_{\alpha}C_{b}$ and $A_{\gamma}C_{\alpha}C_{\beta}$ angles measure 106° + 60° = 166° and the theoretical value is 110.76° + 55.38° = 166.14° (Figures 18 and 19). Thus curved branches built on the 166° angle model and complementary angle are not distorted.

Hypothesis of Pattern Generation in Cienkowskya

Two main hypotheses of pattern formation have been reported among heliozoa: (1) step by step building of Mt microprisms and (2) elaboration from a template provided by the MTOC.

The first attempt to explain axonemal Mt nucleation was made by Roth *et al.* (1970) who suggested that conformational changes induced by bonding acts through several subunits to form an allosteric gradient. One year later, Tilney (1971) proposed that Mt-patterns arise by specific interactions between tubules brought about by bridges, or in other terms, once a Mt has one linker, the other linkers are specified. In a modified hypothesis of the allosteric gradient, Bardele (1977a, b) postulated that the "binding of the linker does not act on the subunits of one Mt only in specifying where the next linker will bind, but that these changes are able to traverse linkers and to affect also neighbouring Mt's". This suggests that the position of both the linkers and the Mt's in the axoneme are defined.

The Central Hexagon

If the central hexagon of *Cienkowskya* is made step by step via the mechanism of the linker nucleation hypothesis with series of equal linkers (7 nm long) to close the hexagon, the last linker on Mt 6 should bind with a site on the Mt 1. But we have seen that because Mt's are composed of 13 subunits, there is a 27.69° angle between the last linker of Mt 6 and the linking site on Mt 1. This difference corresponds to a length of around 14 nm between the end of the linker and the binding site of the Mt 1 and a distance of 14 + 7 = 21 nm between Mt's 1 and 6. One has either to imagine that a drift of 2.3° occurs as soon as the Mt is built or that the distance of 14 nm is compatible with closing, after which simultaneous repartition of the drift is transfered to every angle of the hexagon. This last possibility seems unlikely because the distance of 21 nm seems too long to make linkage possible. Indeed, if we consider the possibilities of linkage in the whole pattern, we can see (Figure 17) that between the Mt A_8 and the inter-row Mt there is no linkage although the measured distance is 13 to 18 nm and both Mt's could accept one linker more. Thus, the distance of 21 nm between the Mt 6 and the Mt 1 of the central hexagon seems *a fortiori* to be too long to permit closing.

An evenly distributed compensation becomes more credible if one refers to a 'linear scaffolding device' present in the centre of each microprism of Mt's (Tilney, 1971; Bardele, 1977a, b). The presence of a dense rod in the centre of each hexagon has been demonstrated with high voltage electron microscopy (HVEM) in *Raphidiophrys* (Rieder, 1979) and has been seen with conventional EM in other species (Febvre-Chevalier, 1973a; Davidson, 1975; Brugerolle and Mignot, 1983a, b). Rieder (1979) considers that this rod, 20 nm in diameter, is involved in initiation and positioning of the Mt-pattern, particularly in 'open and mixed patterns' (see Bardele, 1977a). Although the fixation does not allow us to detect such a feature, it is probably also present in *Cienkowskya*.

The Whole Pattern

The attempt to build the whole pattern using the linker nucleation hypothesis leads us to define at least four different graded conformational changes. The first one, with 11 subunits allows to build the central hexagon (Figure 19). The second, with 6 subunits gives rise to the triangles surrounding the central hexagon. This is in agreement with the scheme of Bardele (1977a, b) for *Acanthocystis* or *Raphidiophrys*. But if we try to close the enneagons we have at least three successive gradients, one with 14 subunits, one with 12 subunits and one with 11 subunits. In order to close the heptagons, gradients with 11 subunits must alternate with gradients with 12 subunits. This succession of gradients is not in agreement with the rule of graded conformational change which predicts that "the closing of a polygon gives rise to the spreading of a new signal".

As discussed earlier, curved branches or arcs (radial branches and series of inter-row Mt's) have a fixed number of Mt's, radial branches are free at the periphery of the pattern i.e. not inter-connected by links, and the basic angle for these branches and series of inter-row Mt's is 166°. It is possible (Figure 19) that the closing of the triangles surrounding the central hexagon gives rise to a gradient of 16 subunits which generates primary branches at 166° angle. The growth of the primary branches (limited to 6 Mt's) may be genetically controlled and could spread the gradient of 6 subunits which form the triangles lying at the origin of the secondary branches. After their closing, secondary and tertiary branches could be generated in the same way. Linkage between branches could be induced by the triangles at the base of the secondary branches with a gradient of 16 subunits. However, such an eventuality is not always possible and one has to admit that lozenges or squares depend on gradient of 9 subunits.

A posteriori explanation of pattern formation, according to the theory of conformational graded change, is tempting and allows the elucidation of rules usable for simple patterns like those of *Heterophrys*, *Acanthocystis* or *Raphidiophrys* but, it is more difficult to explain whole pattern formation in complex cases such as in *Cienkowskya*. However, the progressive increase of the diameter of the axoneme by peripheral adjunction of Mt's supports this hypothesis.

If a linear 'scaffolding device' plays an active part as suggested by Rieder's observations, one would have to imagine in *Cienkowskya* different kinds of generation rods corresponding to the different kinds of polygons. In such a hypothesis, repartition of rods would be controlled by a "pre-existing pattern of nucleating elements or a microtubule nucleating template" as in the ciliate *Nassula* (Tucker, 1970; Tucker *et al.*, 1975). However, in the heliozoans such bonding nucleating elements of the MTOC in a highly ordered spatial array or microtubule nucleating templates have never been found.

4.2. The MTOC

Terminology and Definition

Dobell (1917) described the centroplast as a "clear sphere with a deeply staining granule at its centre in which axes of the axopodia are rooted". The central region of the centrohelidan heliozoa displays variability in size, position and microanatomy in different species. Tilney (1971) described the centroplast in *Raphidiophrys* as a dense sphere including cotton-like material. Bardele (1970, 1975, 1977a, b) observed in various species of *Acanthocystis*, *Heterophrys* and *Raphidiophrys* a central tripartite disc sandwiched between two hemispheres of dense material, both surrounded by a clear 'exclusion zone' and an outer cortex (or 'shell of the centroplast') in which bases of the axonemes were inserted, Davidson (1975) noted "3 central discs surrounded by a dense spherical shell of fibrous material into which the axopods insert. The middle disc is flat with both faces covered with parallel rows of paired helically wound filaments". Rieder (1979) with HVEM described a 'centrosphere' made of three components: "the central tripartite disc, a sphere of electron opaque material which immediately surrounds the central disc, and a slightly less electron opaque shell which consists of components associated with the proximal 1-2 µm of each axonemal base". The succession of the constitutive elements defined above corresponds, with only minor variations, to our observations of *Cienkowskya*. The central disc is trilamellar and embedded in a sphere of dense material as in other species, although its fine organization is not clearly defined and its size is much larger (0.5 µm in *Raphidiophrys*; 1.5 µm in *Cienkowskya*). The exclusion zone, though larger than in other species is similarly made of a fibrillar network connecting the central granule to the base of the axonemes, as previously described by Bardele (1975). In contrast with *Heterophrys marina* and *H. elati*, the bases of the axonemes are not closely connected by a dense material (the cortex of the centroplast according to the nomenclature of Bardele). They remain independent from the neighboring axonemes and are only linked to one another with a thin microfibrillar network (Figures 3, 4, and 14). We are unable to see a difference between the exclusion zone surrounding the central granule and the inter-rod material which extends up to the sheet of cytoplasm filled with organelles and in which the proximal part of the axonemes are rooted. The space between the Mt's (on 5 μm) is packed with a very dense substance.

In order to unify the terminology, we think it is necessary to eliminate the terms shell and cortex of the centroplast which have been used for different parts of the organelle. We have also to recall that there are at least 5 species of heliozoa in which no trace of central granule has been detected. In *Hedraiophrys hovassei*, *Gymnosphaera albida* and *Actinocoryne contractilis* (Febvre-Chevalier, 1973a, 1975, 1980; Jones, 1976), bundles of Mt's arise from a bulky mass of microgranular and microfibrillar material. Very recently, Brugerolle and Mignot (1983a, b) found in the helioflagellate *Dimorpha mutans* an MTOC lying in the centre of the nucleus and in *Tetradimorpha* a lens-like MTOC occupying one pole of the nucleus from which arise axonemal Mt's of the

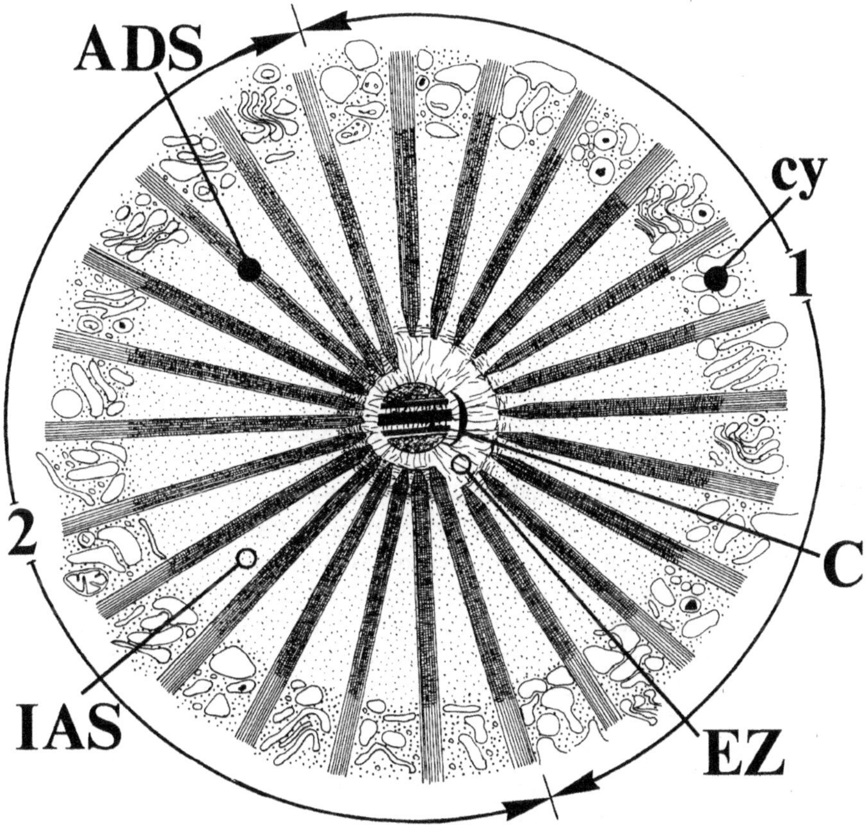

Fig. 20. Interpretative drawing of the MTOC in a Centroplasthelida. Half part 1 corresponds to *Cienkovskya*. Half part 2 is for other genera of Centroplasthelida. C = Centroplast. EZ = Exclusion zone. IAS = Interaxonemal substance. ADS = Axonemal dense substance. cy = cytoplasm.

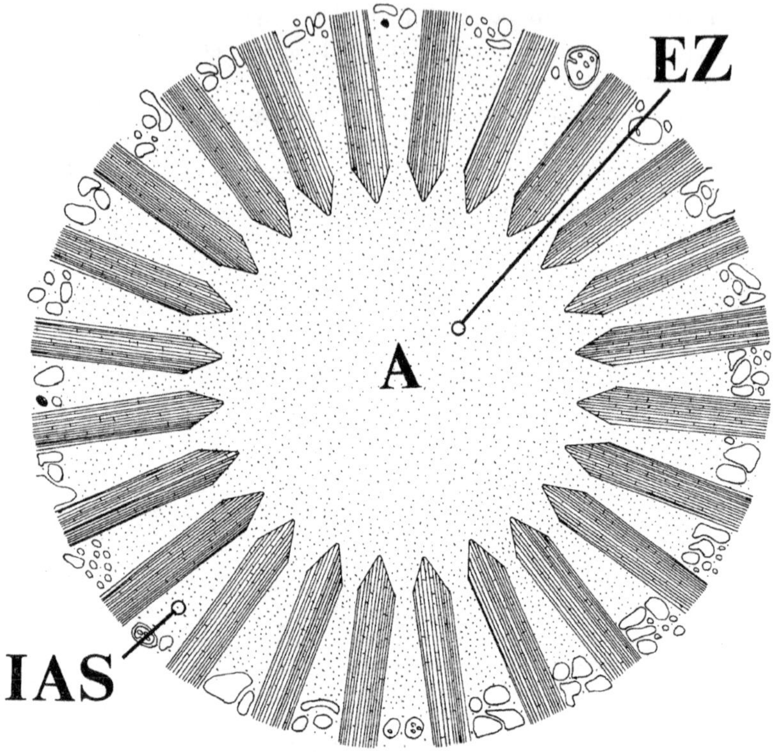

Fig. 21. Axoplast of Axoplasthelida. A = axoplast.

axopods and flagella. These MTOC's are not centroplasts, as claimed by the authors, but axoplasts.

We propose the following definitions which take previous data into account and which use some of the existing terminology (see Figure 20 and 21). An MTOC is defined as the material from which Mt's arise regardless of size, shape or inner substructure. The centroplast (Figure 20) is defined as the central dense sphere composed of a trilamellar disc sandwiched between two spherical caps of electron dense material. The centrosphere is defined as the clear surrounding sphere which is itself divided into two sheets, a clear exclusion zone (EZ) and a clear interaxonemal substance (IAS) in which axonemes are rooted. The axonemal dense substance (ADS) is defined as an electron opaque material which embeds the Mt's at the base of the axonemes in the outer sheet of the centrosphere. The term axoplast (Hollande and Enjumet, 1954; Febvre-Chevalier, 1973a) is reserved for a central or sub-central sphere without centroplast. It can be divided into two regions, (1) a central exclusion zone (EZ) made of microfibrillar and microgranular substance and (2) an inter-axonemal substance (IAS) morphologically similar to the EZ, in which axonemes are inserted (Figure 21). Henceforth we will use the terms as defined above.

Are centroplast, axoplast or centrosphere the MTOC?
On basis of assembly-disassembly experiments, Tilney not observing the tripartite disc, thought that the centroplast was responsible for nucleation of axonemal Mt's. Bardele (1977a) considered that "though experimental proof i.e. by using isolated centroplast in *in vivo* assembly of tubulin is lacking, there can be little doubt that the centroplast serves as MTOC". However, in some species of *Acanthocystis*, the Mt's are directly connected to the centroplast, a fact which seems to weaken the above assumption. However in *Cienkowskya* there is a large exclusion zone, supporting the idea that the centroplast plays a weak part in Mt-assembly. On the other hand, the lack of centroplast in several species of heliozoa (and in other actinopoda) supports the supposition that the whole sphere is the MTOC. The basic unit of the Mt-pattern is nucleated at the outer limit of the exclusion zone in most cases and the pattern extends by progressive addition of Mt's to the periphery of the pattern. This explains the conical shape of the proximal ends of the axonemes.

4.3. Contribution of EM to the Taxonomy of the Heliozoa

Taxonomic Relationships between Cienkowskya and the Other Heliozoa
Cienkowskya mereshkovskyi Cienk., 1881 has a centroplast. Although its axonemal pattern reaches a great complexity it is organized around one central, regularly distorted hexagon. The kinetocysts have a highly organized substructure. Mitochondria have lamellar cristae. All these characters can be seen in *Raphidiophrys*, *Acanthocystis* and *Heterophrys*. The spicules are needle-shaped, cruciform in cross section and probably made of organic substances. They are elaborated in SGO's associated with cell membrane invaginations. These characters with those cited above have been previously reported in different species of *Heterophrys* by Bardele (1975, 1977a) and Davidson (1975) suggesting that both genera are closely related. However, it appears that enneagons, heptagons, squares and lozenges which are delineated by curved branches built from the 166° angle represent a generic character. Therefore, the genus *Cienkowskya* Schaudinn, 1896 is different from the genus *Heterophrys* Archer, 1896.

Hierarchy of Taxonomic Criteria
As early as 1926, Kühn from the observations of Ehrenberg (1830), Penard (1904) and Dobell (1917) claimed that the presence or absence of a central body from which arise axopodial axes was the most important criterion to classify the heliozoa. He distinguished two orders: Actinophryida and Centrohelida, a classification which was later completed but not suppressed (Levine *et al.*, 1980; Febvre-Chevalier, 1982, 1983).

It appears that in Heliozoa as well as in Polycystina (Cachon and Cachon, 1972a, b, 1977) and Acantharia (Cachon *et al.*, 1973), the MTOC and patterned arrays of Mt's are omnipresent and genetically stable. Thus they represent criteria of very high level (Orders) and they display sufficient microanatomical differences to allow characterization of suborders.

Occurence of a stalk seems of minor importance. We think that *Actinocoryne* which is stalked is more closely related to *Gymnosphaera* which is free living than to

Cienkowskya, because *Actinocoryne* has an axoplast like *Gymnosphaera* whereas *Cienkowskya* has a centroplast.

Other criteria such as fine structure of extrusomes, cell coat and skeleton are widely distributed and display sufficient diversity to permit characterization of families. Other characters, for example, number of nuclei and peripheral organization of Mt-patterns are generic or specific criteria.

The presence of flagella (helioflagellates) could be considered as a criterion of very high level permitting the definition of a class: the Helioflagellata with 2 orders, one having a discreet MTOC, mucocysts and flagella with mastigonemes, the other with a conspicuous MTOC, kinetocysts and flagella devoid of mastigonemes. However, we think that present knowledge on helioflagellates is too incomplete to envisage such a rehandling of the taxonomy.

A Revised Classification of the Heliozoa Based on EM Data

Heliozoa have been divided into five orders: Actinophryida Kühn, 1926; Desmothoracida Hertwig and Lesser, 1874; Taxopodida Fol, 1883; Ciliophryida Febvre-Chevalier, 1982 and Centrohelida Kühn, 1926 (and one additional order of questionable validity, the Rotosphaerida Rainer, 1968). The first four orders are characterized by MTOC's consisting of small amounts of dense material either scattered on the nuclear envelope and cytoplasm (Tilney and Porter, 1965; Bardele, 1972; Jones and Tucker, 1981) or looking like simple dense plaques (Cachon and Cachon, 1978; Cachon *et al.*, 1977). These four suborders are also defined by their patterned arrays of Mt's. The fifth order was defined by presence of only one central MTOC with or without inner differentiation. Recent advances in cytology of Heliozoa allow the proposal of a more complete classification with homogeneous terminology.

Conspectus of the Classification

Super-class	ACTINOPODA Calkins, 1909
Class	HELIOZOA Haeckel, 1866
Order I.	CRYPTAXOHELIDA Febvre-Chevalier and Febvre, 1983
Sub-order 1.	Actinophryida Kühn, 1926
2.	Desmothoracida Hertwig and Lesser, 1874
3.	Ciliophryida Febvre-Chevalier, 1982
4.	Taxopodida Fol, 1882
Order II.	PHANERAXOHELIDA Febvre-Chevalier and Febvre, 1983 (ex Centrohelida Kühn, 1926)
Sub-order 5.	Axoplasthelida Febvre-Chevalier and Febvre, 1983
6.	Centroplasthelida Febvre-Chevalier and Febvre, 1983
7.	Endonucleoaxoplasthelida Febvre-Chevalier and Febvre, 1983
8.	Exonucleoaxoplasthelida Febvre-Chevalier and Febvre, 1983

Advantages of This Taxonomy and Nomenclature

Although Levine and coworkers took EM data into account in their taxonomy, they considered groups of heterogeneous features instead of proposing an hierarchy of usable characters. The placed taxa at equivalent levels and did not consider variability of characters. Previously published work (Febvre-Chevalier, 1982) and the key proposed in the present paper allow us to differentiate heliozoan taxa based on variation of only one pair of characters: MTOC and Mt-patterns. While in our previous work only orders were characterized, the data given in this paper confirm the variability of this pair of taxonomic criteria and allow the definition of orders and suborders. Families are further defined by microanatomical variation of extrusomes, cell coat and spicules.

In our key, new names, derived from Greek roots, are proposed which reflect differences in the primary criterion (MTOC) except for the 4 first sub-orders whose names remain unchanged. Actinophryida, Desmothoracida, Ciliophryida and Taxopodida have imperceptible MTOC's in the light microscope and have been regrouped in the order Cryptaxohelida (crypto = kruptos = hidden; heliozoa with hidden MTOC). The name centrohelida has been replaced by Phaneraxohelida, the prefix phanero (phaneros = conspicuous) counterbalances crypto. The new name Phaneraxohelida is more suitable because it avoids misunderstanding with the names proposed for the 4 subsequent suborders. Axoplasthelida has been chosen for Phaneraxohelida with a central axoplast while Centroplasthelida designates Phaneraxohelida with a centroplast. Endonucleoaxoplasthelida, one order of phaneraxohelida, includes those species with an axoplast surrounded by the nucleus. Exonucleoaxoplasthelida is used for Phaneraxohelida in which the MTOC is an axoplast located outside of the nucleus. Taxonomic criteria in the class Heliozoa are similar to those previously adopted in the Polycystina making homogeneous the nomenclature in both classes of Actinopods.

5. Class Heliozoa

General Characters of the Heliozoa from L.M. Data

Heliozoa are Actinopods 5 to 500 µm in diameter with long, radiating, more or less contractile, granule-studded axopods stiffened by an axial rod. Axopods sometimes displaying rowing motion. In some cases one or more flagella in addition to axopods. Axopodial axis arising from one central or subcentral centre or from the nuclei. No central capsule. Cell naked or covered with an outer dispersed skeleton. Siliceous or organic spicules. With or without tangential scales. Some species with an organic perforated shell. With or without conspicuous mucous coat. One or several nuclei. Generally heterotrophic nutrition. Some with symbiotic algae. Free-living or sessile. Some with alternation between spherical and stalked forms. Stalk either mucous, or cytoplasmic and highly contractile. Asexual reproduction by binary or multiple fission or budding. Sexual reproduction by autogamy (pedogamy) or biflagellate swarmers. Cysts = gamontocysts or resting cysts.

Characters of the Heliozoa from E.M. Data
Fine structure of the axonemes consisting of highly organized patterns of microtubules. Open, mixed or closed patterns. MTOC's consisting of small amounts of dense material generally dispersed on the surface of the nuclear envelope and not very conspicuous, or MTOC's consisting of a central conspicuous sphere of subcentral lens including one centroplast or made of an axoplast. Extrusomes or extrusive organelles elaborated from the golgi system consisting of dense granules, fibrous-core granules, mucocysts or kinetocysts. Mucous coat without particular geometrical arrangement. Skeleton made of hollow or full spicules issuing from the golgi system or secreted in spicule generating organelles (SGO). Mitochondria with tubular or ampula-like, or vermicular or lamellar cristae.

6. Key to the Heliozoa from E.M. Data

– MTOC discernible with E.M. only. Generally scattered on the nuclear envelope and consisting of simple or stratified dense material.

Order CRYPTAXOHELIDA I.

– Only one bulky MTOC located at the centre of the cell, with or without a central differentiation.

Order PHANERAXOHELIDA II.

I. Order CRYPTAXOHELIDA
– MTOC's scattered on the outer nuclear membrane or within the cytoplasm. Interlocking coils of axopodial Mt's.

Sub-order *Actinophryida* 1.

– MTOC's on the outer nuclear membrane. Imprecise Mt-pattern.

Sub-order *Desmothoracida* 2.

– MTOC's are simple dense plaques on the nuclear envelope. Pattern of 3 Mt's in triangle. Both flagellate and heliozoa forms present during the vegetative life.

Sub-Order *Ciliophryida* 3.

– MTOC's are stratified dense plaques articulated in caveolae of the nuclear envelope. Irregular hexagonal flooring pattern.

Sub-order *Taxopodida* 4.

II. Order PHANERAXOHELIDA
– MTOC's are central spherical axoplasts. Irregular hexagonal Mt-flooring pattern.

Sub-Order *Axoplasthelida* 5.

– MTOC with a centroplast. Central basic Mt-pattern composed of one distorted hexagon with 106° and 134° angles.

Sub-Order *Centroplasthelida* 6.

– MTOC is an axoplast surrounded by the nucleus. Quincunx pattern of Mt's.

Sub-Order *Endonucleoaxoplasthelida* 7.

– MTOC is a lens-shaped axoplast lying externally to the nucleus. Imprecisely defined Mt-pattern.

Sub-Order *Exonucleoaxoplasthelida* 8.

I. O. CRYPTAXOHELIDA

1. Sub-Order *Actinophryida*

Extrusomes are dense granules. Naked. One or several nuclei.

F. **Actinophryidae**

– 1 nucleus — G. *Actinophrys*

– several nuclei — G. *Actinosphaerium*

2. Sub-Order *Desmothoracida*

– Extrusomes are kinetocysts. Organic latticed capsule.

F. **Clathrulinidae**

– 1 nucleus — G. *Clathrulina*

3. Sub-Order *Ciliophryida*

– Extrusomes are mucocysts. Naked. Mastigonemes on the flagellum.

F. **Ciliophryidae**

– 1 nucleus — G. *Ciliophrys*

4. Sub-Order *Taxopodida*

– No extrusomes. Thin elastic coat. Long siliceous hollow radial spatula-like spicules and curved short spicules.

F. **Sticholonchidae**

– 1 nucleus with a thick nuclear envelope. Rowing motion of the axonemes by means of bundles of twisted microfilaments.

G. *Sticholonche*

II. O. PHANERAXOHELIDA

5. Sub-Order *Axoplasthelida*

– Extrusomes including dense granules + mucocysts. No skeleton. Thin cell coat. One or several nuclei.

F. **Gymnidae** Febvre-Chevalier and Febvre, 1983

– Several nuclei. No sessile stage. G. *Gymnosphaera*

– Young free-living. 1 nucleus.
Adult sessile, several nuclei. Contractile stalk with large bundles of Mt's
G. *Actinocoryne*

– Extrusomes are dense oval granules + mucocysts.Thick coat. Siliceous spicules.
F. **Hedraiophryidae**

– One large nucleus. Adult free living or sessile.
G. *Hedraiophrys*

6. Sub-Order *Centroplasthelida*

– Extrusomes are kinetocysts or kinetocysts and mucocysts. Thin or thick skeletal capsule. Organic spicules, cruciform in cross section, made in generating organelles (SGO). With or without mucous stalk.
F. **Heterophryidae**

– Extrusomes are kinetocysts. Thin capsule. Limited number of axonemal Mt's. No mucous stalk.
G. *Heterophrys*

– Extrusomes are kinetocysts + mucocysts. Thick capsule consisting of cell coat and spicules. Hollow mucous stalk. Large Mt-pattern with peripheral different kinds of polygons.
G. *Cienkowskya*

– Siliceous scales + spicules. Extrusomes are kinetocysts formed in golgi system.
F. **Acanthocystidae**
G. *Acanthocystis*

– Extrusomes are kinetocysts. Siliceous radial spicules.
F. **Raphidiophryidae**
G. *Raphidiophrys*

7. Sub-Order *Endonucleoaxoplasthelida*

– Extrusomes are kinetocysts with cylindrical substructure. 2 naked flagella orthogonally disposed.
F. **Dimorphyidae**

– 1 nucleus G. *Dimorpha*

8. Sub-Order *Exonucleoaxoplasthelida*

– Extrusomes are kinetocysts. 4 naked flagella anteriorly oriented.
F. **Tetradimorphyidae**

– 1 nucleus G. *Tetradimorpha*

Acknowledgments

It is our pleasure to acknowledge Pr. Christian Bardele for reading the manuscript and helpful comments and Dr. David Brownlee for revising the English text.

References

Bardele, C. F.: 1970, Comparative Ultrastructural Studies on Centrohelida, *J. Protozool. Suppl.* **17**.

Bardele, C. F.: 1972, Cell Cycle, Morphogenesis and Ultrastructure in the Pseudoheliozoan *Clathrulina elegans*, *Z. Zellforsch.* **130**, 219–242.

Bardele, C. F.: 1975, The Fine Structure of the Centrohelidan Heliozoan *Heterophrys marina*, *Cell Tiss. Res.* **161**, 85–102.

Bardele, C. F.: 1977a, Comparative Study of Axopodial Microtubule Patterns and Possible Mechanisms of Pattern Control in the Centrohelidian Heliozoa *Acanthocystis*, *Raphidiophrys* and *Heterophrys*, *J. Cell Sci.* **25**, 205–232.

Bardele, C. F.: 1977b, Organization and Control of Microtubule Pattern in Centrohelidan Heliozoa, *J. Protozool.* **24**, (1), 9–14.

Brugerolle, G. and Mignot, J. P.: 1983a, Etude Ultrastructurale de l'Hélioflagellé *Dimorpha mutans*. 22ème Réunion du G.P.L.F. Murat le Quaire, Juin 1983.

Brugerolle, G. and Mignot, J. P.: 1983b, Fine Structure of the Helioflagellates *Dimorpha* and *Tetradimorpha*, a Contribution to the Knowledge of the Evolution in Heliozoa. 5th. Meet. Intern. Soc. Evol. Protistology, Banyuls-sur-Mer (France), Juin 1983.

Cachon, J. and Cachon, M.: 1972a, Le système Axopodial des Radiolaires Sphaeroidés. I. Centro-axoplastidiés. *Arch. Protistenk.* **114**, 51–64.

Cachon, J. and Cachon, M.: 1972b, Le Système Axopodial des Radiolaires Sphaeroidés. II. Les Périaxoplastidiés. III. Les Cryptoaxoplastidiés (Anaxoplastidiés). IV. Les Fusules et le Système Rhéoplasmique, *Arch. Protistenk.* **114**, 291–307.

Cachon, J. and Cachon, M.: 1974, Les Systèmes Axopodiaux, *Ann. Biol.* **13**, 523–555.

Cachon, J. and Cachon, M.: 1977, Le Système Axopodial des Collodaires (Radiolaires Polycystines). 2. *Thalassolampe margarodes* Haeck, *Arch. Protistenk.* **119**, 401–406.

Cachon, J. and Cachon, M.: 1978, *Sticholonche zanclea* Hertwig. A Reinterpretation of its Phylogenetic Position Based Upon New Observations on Its Ultrastructure, *Arch. Protistenk.* **120**, 148–168.

Cachon, J., Cachon, M., Febvre-Chevalier, C., and Febvre, J.: 1973, Déterminisme de l'Edification des Systèmes Microtubulaires Stéréoplasmiques d'Actinopodes, *Arch. Protistenk.* **115**, 137–153.

Cachon, J., Cachon, M., Tilney, L. G., and Tilney, M.: 1977, Movement Generated by Interaction Between the Dense Material at the End of Microtubules and Non Actin Containing Microfilaments in *Sticholonche zanclea*, *J. Cell Biol.* **72**, 314–338.

Cienkovsky, L.: 1881, Bericht über die Excursion ins Weisse Meer d.J. 1880. Arbeiten den St. Petersb, *Natürf. Gesellsch.* **12**, 130–171.

Davidson, L. A.: 1972, Ultrastructure of the Heliozoan *Heterophrys marina* and of the Helioflagellate *Ciliophrys marina*, *J. Ultrastr. Res.* **38**, 213–222.

Davidson, L. A.: 1974, An Outline of the Origine and Evolution of the Actino pods (Abstr.), *J. Protozool.* **21**, 426.

Davidson, L. A.: 1975, Studies of the Actinopods *Heterophrys marina* and *Ciliophrys marina*: Energetics and Structural Analysis of Their Contractile Axopodia, General Ultrastructure and Phylogenetic Relationships. PhD. Thesis, University of California at Berkeley.

Davidson, L. A.: 1976, Ultrastructure of Membrane attachment Sites of the Extrusomes of *Ciliophrys marina* and *Heterophrys marina* (Actinopoda), *Cell Tiss. Res.* **170**, 353–365.

Davidson, L. A.: 1982, Ultrastructure, Behavior and Algal Flagellate Affinities of the Helioflagellate *Ciliophrys marina* and the Classification of the Helioflagellates (Protista-Actinopoda-Heliozoea), *J. Protozool* **29**, 19–29.

Dobell, C.: 1917, On *Oxnerella maritima* nov. gen. nov. spec. a New Heliozoon and Its Method of Division, with Some Remarks on the Centroplast of the Heliozoa, *Quart. J. Microsc. Sci.* **62**, 515–538.

Dustin, P.: 1978, *Microtubules*, Springer-Verlag, Berlin, Heidelberg, New York, pp. 1–452.

Ehrenberg, C. G.: 1830, *Actinophrys sol in*: Beiträge zur Kentniss der Organisation der Infusiorien und ihrer geographischen Verbreitung . . ., *Abhandl. König. Acad. Wiss. Berlin.* 1–88.

Febvre-Chevalier, C.: 1973a, *Hedraiophrys hovassei* nov. gen. nov. sp. Morphologie biologie et cytologie, *Protistologica* **9**, 503–520.

Febvre-Chevalier, C.: 1973b, Un Nouveau Type d'Association des Microtubules Axopodiaux chez les Héliozoaires, *Protistologica* **9**, 35–43.

Febvre-Chevalier, C.: 1975, Etude Cytologique de *Gymnosphaera albida*, Sassaki 1894. (Héliozoaire Centrohélidié), *Protistologica* **11**, 331–344.

Febvre-Chevalier, C.: 1980, Behaviour and Cytology of *Actinocoryne contractilis*, nov. gen. nov. sp. A New Stalked Heliozoan (Centrohelidia). Comparison with the Other Related Genera, *J. Mar. Biol. Ass. U.K.* **60**, 909–928.

Febvre-Chevalier, C.: 1981, Preliminary Study of the Motility Processes in the Stalked Heliozoan *Actinocoryne contractilis*, *BioSystems* **14**, 337–343.

Febvre-Chevalier, C.: 1982, Révision of the Taxonomy of the Heliozoa with Attention to Electron Microscopical Criteria, *Ann. Inst. Oceanogr. Paris* **58**, (S) 173–178.

Febvre-Chevalier, C.: 1983, Class III. Heliozoa, in J. J. Lee, S. H. Hutner and E. C. Bovee (Eds.), *Illustrated Guide to the Protozoa*, pp. 15.

Febvre-Chevalier, C. and Febvre, J.: 1980, Cytophysiologie de la Motilité chez un Héliozoaire Pédonculé. Film SFRS, Paris.

Hollande, A. and Enjumet, M.: 1954, Sur l'Existence d'Axopodes et d'un Complexe Centroplastique chez les Radiolaires. *C.R. Acad. Sci., Paris* **238**, 1841–1843.

Jones, W. C.: 1976, The Ultrastructure of *Gymnosphaera albida* Sassaki, a Marine Axopodiate Protozoon, *Phil. Trans. Roy. Soc. London* **275**, 349–384.

Jones, C. and Tucker, J. B.: 1981, Microtubule Organizing Centres and Assembly of the Double Spiral Microtubule Pattern in Certain Heliozoan Axonemes, *J. Cell Sci.* **50**, 259–280.

Karnovsky, M. J.: 1965, A Formaldehyde-Glutaraldehyde Fixative of High Osmolarity for Use in Electron Microscopy, *J. Cell Biol.* **27**, 137 A.

Kühn, A.: 1926, Morphologie der Tiere in Bildern 2 H Protozoen. 2. Teil Rhizopoden. Berlin, 187–272.

Levine, N. D., Corliss, J. O., Cox, F. E. G., Deroux, G., Honiberg, B. M., Leedale, G. F., Loeblich, A. R. III, Lom, J., Lynn, D., Merinfeld, E., Page, F. C., Poljansky, G., Sprague, V., Vavra, J., Wallace, F. G.: 1980, A Newly Revised Classification of the Protozoa, *J. Protozool.* **27**, 37–58.

Penard, E.: 1904, Les Héliozoaires d'Eau Douce, Genève, 1–341.

Pickett-Heaps, J. D.: 1969, The Evolution of the Mitotic Apparatus. An Attempt at Comparative Ultrastructural Cytology in Dividing Plant Cells, *Cytobios.* **1**, 257–280.

Rieder, C. L., 1979, High Voltage Electron Microscopy Reveals New Components in the Axoneme and Centroplast of the Centrohelidian *Raphidiophrys ambigua*, *J. Cell Sci.* **40**, 215–234.

Roth, L. E., Pihlaja, D. J., and Shigenaka, Y.: 1970, Microtubules in the Heliozoan Axopodium I. The Gradion Hypothesis of Allosterism in Structural Proteins, *J. Ultrastr. Res.* **30**, 7–37.

Spurr, A.: 1969, A Low Viscosity Epoxy Resin Embedding Medium for Electron Microscopy, *J. Ultrastr. Res.* **26**, 31–43.

Tilney, L. G.: 1971, How Microtubule Patterns Are Generated? The Relative Importance of Nucleation and Bridging of Microtubules in the Formation of the Axoneme of *Raphidiophrys*, *J. Cell Biol.* **51**, 837–854.

Tilney, L. G. and Porter, K. R.: 1965, Studies on Microtubules in Heliozoa. I. The Fine Structure of *Actinosphaerium nucleofilum* (Barrett) with Particular Reference to the Axial Rod Structure, *Protoplasma*, **60**, 317–344.

Tucker, J. B.: 1970, Initiation and Differentiation of Microtubule Patterns in the Ciliate *Nassula*, *J. Cell Sci.* 7, 793–821.

Tucker, J. B., Dunn, M., and Pattisson, J. B.: 1975, Control of Microtubule Pattern During the Development of a Large Organelle in the Ciliate *Nassula*, *Develop. Biol.* **47**, 439–453.

Villeneuve, F.: 1937, Sur la Structure de *Cienkovskya mereshkovskyi* Cienk. et d'*Actinolophus pedunculatus* Schulze, Héliozoaires des eaux saumâtres de Sète, *Arch. Zool. Exper. et Gen.* **78**, 243–250.

Zuelzer, M.: 1909, Bau und Entwicklung von *Wagnerella borealis* Mereschk., *Arch. f. Prot.* **17**, 135–202.

THE PHYLOGENY OF THE GREGARINIA (SPOROZOA)

JEAN THÉODORIDÈS

Laboratoire d'Evolution, Université Pierre et Marie Curie, 105 boulevard Raspail, 75006 Paris, France

(Received 19 October, 1983)

Abstract. The Gregarinia, all parasitic in invertebrates have appeared in marine hosts (Polychaeta). The most primitive are the Archigregarinida (*Selenidium*) which are intestinal parasites. The Eugregarinida (intestinal or coelomic) are found in various marine invertebrates as well as in terrestrial or freshwater arthropods. The Neogregarinida which parasitize only insects have a very reduced number of spores compensated by a schizogony (merogony). The Gregarinia constitute a perfect example of a concomitant evolution between a group of parasites and their hosts.

It is now currently admitted that the Sporozoa (Gregarinia, Coccidia, Piroplasma) most probably derive from the Flagellata.

The Gregarinia are generally characterized by the extracellular development of their gamonts which produce the same number of gametes in both sexes. This gamogonic process separates them from the Coccidia in which the female gamont transforms into a single macrogamete. After fecundation sporogony occurs. There is always a great number of spores* in the archigregarines and eugregarines and a very reduced one in neogregarines. These spores contain eight zoites in eugregarines and neogregarines, generally four zoites in the primitive archigregarines.

The extracellular development of most gregarines results from the general tendency of the sporozoites to adhere at the surface of the host tissues (epithelium of the intestinal tract) in which they develop a fixative apparatus called a mucron or epimerite according to its development.

The Gregarinia are all exclusive parasites of invertebrates (marine, freshwater and terrestrial). The most primitive forms appeared in marine hosts.

As it is the rule in all the groups of parasitic organisms, some of them have been associated with their hosts for a long geologic time (phylogenetic parasitism) whereas others infested them later (ethologic or ecologic parasitism).

In a given host or group of hosts one frequently observes both phylogenetic (= ancient) and ecologic (= recent) parasites. Of course the first ones are more strictly associated with their hosts (stenoxeny) than the second which are euryxenous.

The most primitive Gregarinia are the archigregarines of Grassé (1953) (order Archigregarinida in the last classification of Levine *et al.*, 1980) in which the ultrastructure resembles that of the infective stage of Sporozoa (Schrével, 1971).

* We prefer this term to the ones of 'oocyst' or 'sporocyst' which correspond to peculiar coccidian gamogonic and sporogonic stages.

Origins of Life **13** (1984) 339-342. 0302-1688/84/0133-0339$00.60.

In spite of earlier suppositions no recent worker has been able to observe a vegetative merogony in this group. These gregarines are restricted almost exclusively to the Polychaeta Sedentaria, although some species are parasitic in Enteropneusta. The most important genus is *Selenidium* of which the young trophozoites are fixed to the epithelial cells of the host's intestine by their expanded apical part which does not however penetrate inside the cells and is therefore very little differentiated.**

The Lecudinidae (*Lecudina, Lankesteria*, etc.) which parasitize Polychaeta (chiefly Errantia) and also other groups of marine invertebrates (Echiurida, Sipunculida, Tunicata) belong to the order Eugregarinida (eugregarines). Here, the apical part of the young trophozoites develops into a frontal organelle, the mucron, which has a flattened membraneous surface adhering to the host's intestinal cells. In some species, the mucron is situated at the top of a 'trunk' resulting from the elongation of the anterior part of the gregarine. In some genera such as *Metamera* (in marine leeches) there is however a real epimerite with hooks.

In the family Urosporidae (*Gonospora, Cystobia, Pterospora, Diplauxis*) also parasitizing Polychaeta as well as Mollusca and Echinodermata, the young forms (zoites) cross the intestinal wall and reach the body cavity where they finish developing. There is no doubt that there is a further degree of specialization in the association with the host. It is also to be noted that in some genera of the Urosporidae (*Diplauxis, Pterospora*) syzygy (pairing) of gamonts takes place very early.

In the Eugregarinida of marine and freshwater Crustacea (Ganymedidae, Uradiophoridae, Porosporidae, Cephalolobidae, Cephaloidophoridae) there is also early association between trophozoites, but they are all intestinal parasites. In the Cephalolobidae, the anterior part of the trophozoite is differentiated into a 'proto-epimerite' with 'sucker-like' lobes. In the Uradiophoridae, Porosporidae, Ganymedidae and Cephaloidophoridae there is a rudimentary mucron-like epimerite.

The most primitive Eugregarinida in terrestrial hosts seem to be the Monocystidae and Diplocystidae, coelomic parasites of oligochaetes and insects respectively. In these there is no real epimerite*** whereas this organelle is present in the Gregarinidae all parasitic in insects. It is absent in the Enterocystidae which parasitize only freshwater insects (Ephemeroptera larvae)****.

A further specialization in the differentiation of the epimerite as an organ of fixation is the development of hooks or harpoons; this occurs in the Stylocephalidae (exclusive parasites of tenebrionid beetles) and in the Actinocephalidae (found in both insects and arachnids).

** The direct ancestors of the Archigregarinida seem to be the blastogregarines (and more especially the genus *Siedleckia* Chatton et Villeneuve which parasitize the same hosts (Polychaeta). As in Gregarinia, equal number of gametes of both sexes are produced but the gamogony is not preceded by the pairing of gamonts observed in all gregarines. The gametes of *Siedleckia* are the result of the fragmentation of the growing zoites attached to the epithelium of the host's intestine. These gametes are released in the lumen of the gut where fecundation occurs. Ten to sixteen zoites originate from the zygote. They are surrounded by an envelope. The addition of an order Blastogregarinida in the class Gregarinia may be considered.

*** The occurrence of a sort of 'trunk' in *Rhynchocystis* has led some authors to see an unproven relation between the intestinal Lecudinidae and the coelomic Monocystidae.

**** The difference between septate and aseptate Eugregarinida has no significance at the level of the class.

In Eugregarinida of Myriapoda there is a difference between the parasites of Diplopoda (Stenophoridae, Monoductidae) and Chilopoda (Dactylophoridae). In the former, the organ of fixation is very reduced, whereas in the latter there is a complex rhizoid system which penetrates between the host's intestinal cells.

The third order of Gregarinia is the Neogregarinida, all developing in narrow cavities (fat body, malpighian tubules) of terrestrial and freshwater insects. In this group there is a conspicuous merogony giving rise to a great number of meronts and

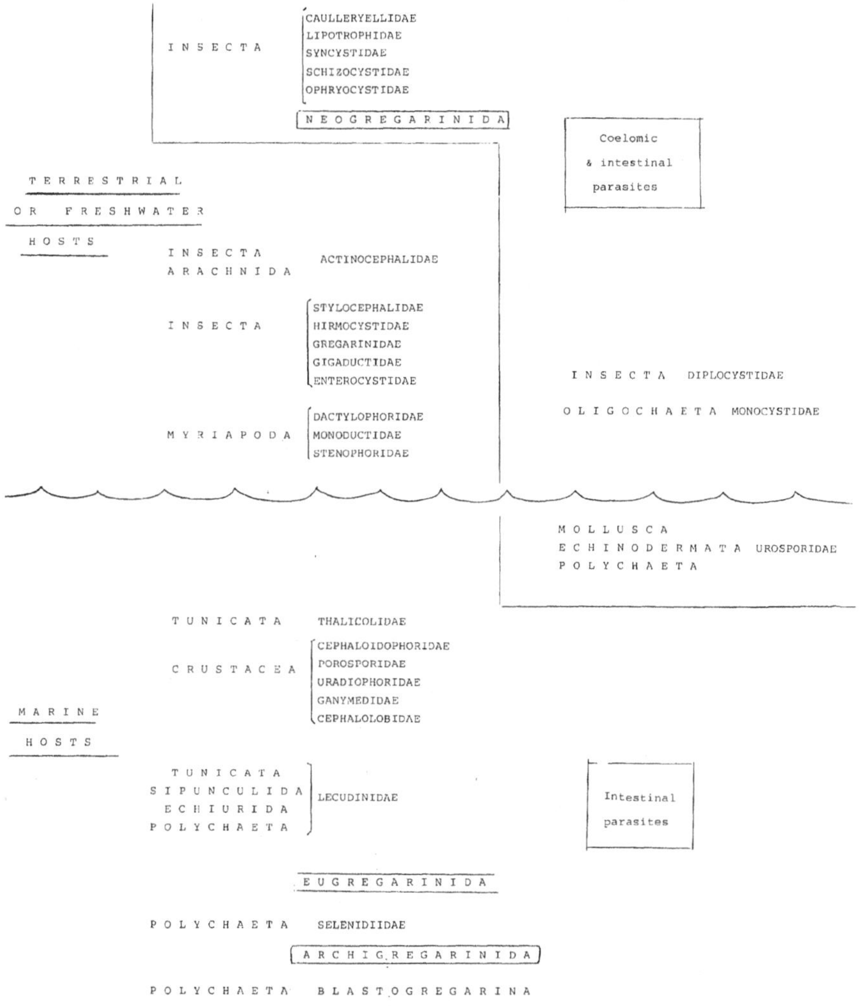

Chart. The main families of Gregarinia arranged according to their hosts. All the families are reported in Grassé (1953) except the Thalicolidae and Cephalolobidae respectively created by Théodoridès and Desportes (1968, 1975).

merozoites which makes these parasites very pathogenic for their hosts. This process compensates for their very reduced sexual reproduction (from one or two to ca. twenty spores).

The Chart indicates that the Gregarinia form a perfect example of concomitant evolution of a group of parasites within their hosts and with the latter.

The group has appeared in primitive marine invertebrates (Polychaeta) from which it has spread in other marine groups (Echiurida, Sipunculida, Mollusca, Echinodermata, Tunicata, Crustacea).

These parasites have then infested terrestrial or freshwater arthropods (Crustacea, Arachnida, Myriapoda, Insecta) in which they have acquired their greatest differentiation. Most of the Gregarinia are intestinal parasites. There are, however, some groups that are exclusively coelomic in marine (Urosporidae) and terrestrial (Monocystidae, Diplocystidae) hosts. Most members of the order Neogregarinida are also coelomic, parasitizing terrestrial and freshwater insects.

In the case of phylogenetic parasitism, specificity is so strict that each species of a given host or group of hosts may harbour its own gregarines. Sometimes one can even identify the host species from its parasite.

References

Grassé, P.: 1953, Grégarinomorphes *in Traité de Zoologie*, Masson, Paris, **1**, pp. 550–690.
Levine, N. D. *et al.*: 1980, *J. Protozool.* **23**, 64–74.
Schrével, J.: 1971, *J. Protozool.* **18**, 448–470.
Théodoridès, J. and Desportes, I.: 1968, *Bull. Inst. Océan. Monaco*, **67**, no. 1387, 1–11.
Théodoridès, J. and Desportes, I.: 1975, *Protistologica*, **11**, 205–220.

THE PARAMYXEA LEVINE 1979: AN ORIGINAL EXAMPLE OF EVOLUTION TOWARDS MULTICELLULARITY

ISABELLE DESPORTES

Cytophysiologie des Protozoaires, 105 bd Raspail, 75006 Paris, France

(Received 12 December, 1983)

Abstract. The Paramyxea are parasitic in marine invertebrates. Their development is a sporulation involving the differentiation within a stem cell of several sporonts which produce spores made of cells enclosed inside each other. Three genera are recognized according to the number of spores and sporal cells, and the taxonomic position of the host (Polychaeta, Mollusca, Crustacea).

The Paramyxea exhibit both protistan and metazoan characters. Their nine singlets centrioles are observed in different Protoctists whereas the fact that their sporal cells acquire distinctive cytological features may be interpreted as an evolution towards multicellularity.

1. Introduction

The Paramyxea are primitive parasites of different marine invertebrates (Polychaeta, Crustacea, Mollusca). *Marteilia refringens* (Grizel *et al.*, 1974) caused very high mortality in the population of the European flat oyster (*Ostrea edulis*). Its sporulation was the subject of an ultrastructural study (Perkins, 1976). *Marteilia sydneyi* is parasitic in Australian oysters (*Crassostrea commercialis*) (Perkins and Wolf, 1976). The development of these parasites starts with a stem cell in which are differentiated several secondary cells. The latter produce spores consisting of several cells enclosed inside each other.

A similar parasite was found in the Amphipod Crustacea *Orchestia gammarellus* (Ginsburger-Vogel *et al.*, 1976). It is considered to be a new genus *Paramarteilia orchestiae* Ginsburger-Vogel and Desportes (1979) due to the fewer number of differentiated spores in the secondary cells and the fewer number of spore cells.

The taxonomic affinities of these parasites were not, however, clearly established. They were first related to Haplosporea (Perkins, 1976) due to the presence of membrane bounded granules similar to the so-called haplosporosomes observed in the latter. All these parasites were later placed in the new class Stellatosporea (Sprague, 1968) (see the new classification of Protozoa by the Committee on Systematics, 1980).

Recently, an ultrastructural study of *Paramyxa paradoxa* first observed by Chatton (1911) in Polychaeta larvae, has suggested that *Marteilia* and *Paramarteilia* are related to this parasite (Desportes and Lom, 1980). They all have been included in the class Paramyxea Levine, 1979 (Desportes, 1981).

2. The Paramyxean Life-Cycle

Two successive sporulation steps are distinguished in the developmental cycle of Paramyxea.

Origins of Life **13** (1984) 343-352. 0302-1688/84/0133-0343$01.50.

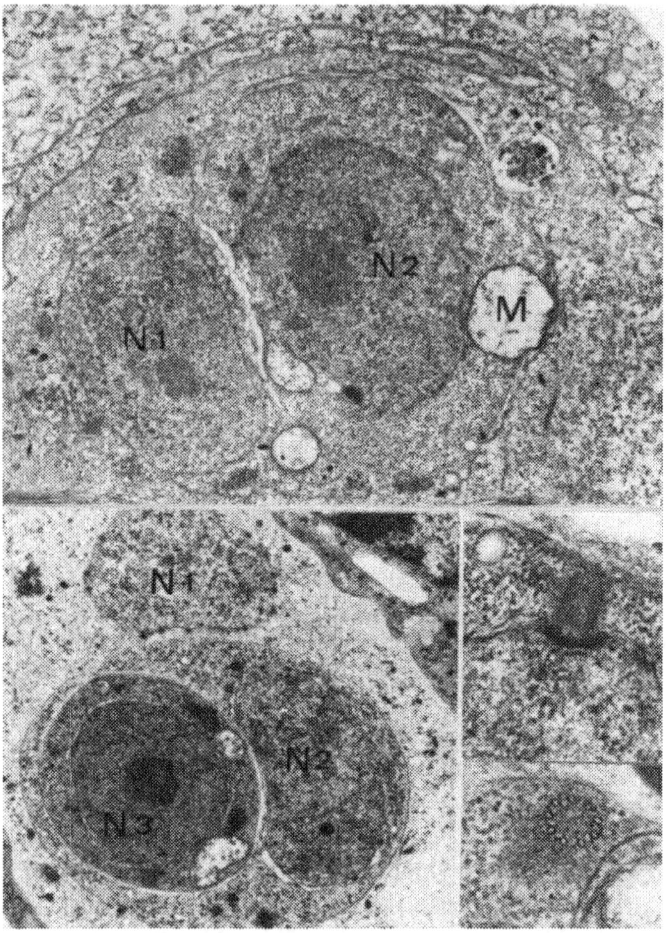

Fig. 1. The differentiation of the first secondary cell within the primary stem cell of *Paramarteilia orchestiae*. N1, N2, nuclei of these cells; M, mitochondrion. × 14 000.

Fig. 2. Budding of the third cell within a secondary one in *Paramarteilia orchestiae*. N1, N2, N3, nuclei of the primary, secondary and tertiary cells. × 9 000 (after Ginsburger-Vogel *et al.*, 1976).

Fig. 3. The polar centriole attached to the nuclear envelope in *P. orchestiae*. × 38 000 (after Ginsburger-Vogel *et al.*, 1976).

Fig. 4. Transverse section of the nine singlets centriole. × 6 000 (after Ginsburger-Vogel and Desportes, 1979).

1. The differentiation of the secondary cells within the primary stem cell.

The first part of the development is similar in all the known members of the group. The young stem cell is amoeboid at the beginning of the life-cycle. It does not penetrate into the host cells but develops between them. Its development is characterized by the continuous enlarging of its cytoplasm. The first nuclear division is most interesting. The stem cell nucleus produces a second nucleus which is at first smaller and surrounded by a thin cytoplasmic layer (Figure 1). The production of this new nucleus results from a mitotic process similar at the beginning to that observed in a chytrid (see Desportes and Ginsburger-Vogel, 1977). We observe (1) the presence at the mitotic poles of two centrioles in an orthogonal arrangement, one of them attached to the nuclear envelope (Figure 3), (2) the differentiation of a spindle with microtubular bundles connected to kinetochores (Figure 5), (3) the subsequent opening of the nuclear envelope.

The rupture of the nuclear envelope is however more important than that observed in Chytrids and this process is most significant in the paramyxean mitosis. Effectively, the envelope is disrupted into vesicles similar to those of the endoplasmic reticulum. All these vesicles by fusing end-to-end complete the envelope of the two nuclei and furthermore, form the pair of membranes delimitating the cytoplasmic layer around the new nucleus.

The so-formed secondary cell divides again. This division differs from the first in that it produces two identical cells. The repetition of the latter process results in the production of a variable number of secondary cells which are characterized by their

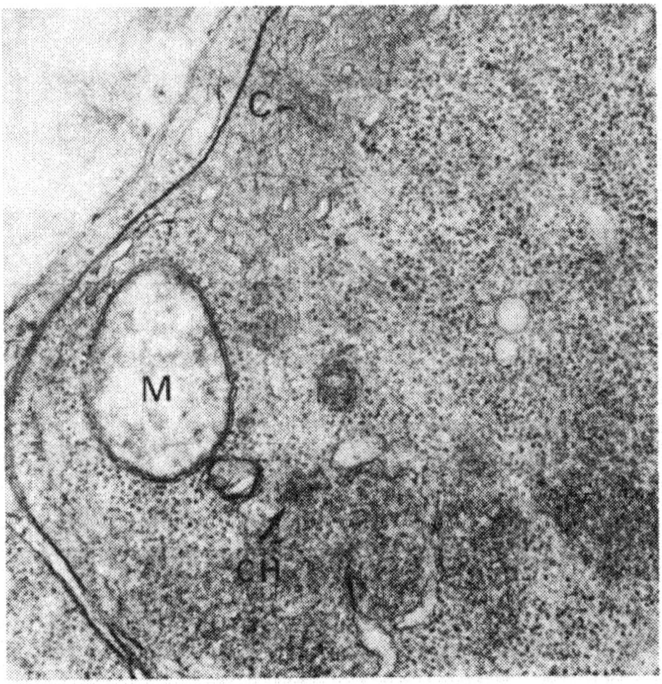

Fig. 5. A dividing secondary cell. The nuclear envelope is disrupted into vesicular fragments visible at the mitotic pole (C, centriole) and around the chromosomes (CH); a kinetochore is indicated by the arrow; M., mitochondrion. × 38 000.

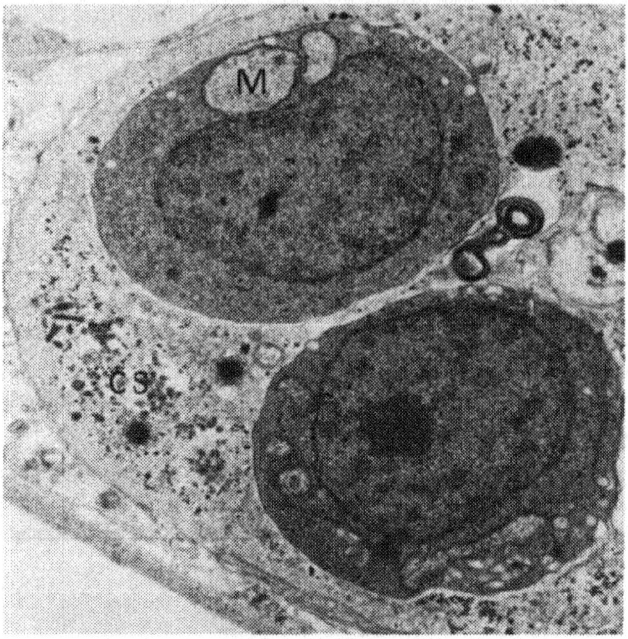

Fig. 6. Two secondary cells in the stem cell cytoplasm (CS) of *P. orchestiae*. × 17 500.

cytoplasm containing numerous membrane-free ribosomes, an electron-dense nucleus and light-matrix mitochondria (Figure 6).

2. The differentiation of the spores within the secondary cells.

The sporulation itself starts with the 'endogenous budding' of a third cell within each secondary cell (Figure 2). At this stage which is also similar in all Paramyxea, the secondary cell may be considered as a 'sporont' within which the third cell undergoes the divisions giving rise to the spores. The number of spores and sporal cells differentiated in the sporonts as well as the taxonomic position of the hosts are the characters distinguishing the three paramyxean genera: *Paramarteilia*, *Marteilia* and *Paramyxa*. The comparative development of their sporonts is summarized in Figure 7.

The division of the third cell produces two primary (= outer) sporal cells in *Paramarteilia* and four primary sporal cells in the other genera. In *Paramarteilia* each primary sporal cell produces a small inner cell which is considered as the sporoplasm (Figures 9 and 10). In *Marteilia* one secondary sporal cell is differentiated between the outer primary sporal cell and the inner sporoplasm (Figure 7) whereas one secondary cell and a tertiary one are differentiated in *Paramyxa* (Figures 7 and 11). Furthermore, an additional and early rejected cell is budded from the tertiary cell, just before the production of the sporoplasm (Desportes, 1981). This cell consisting of an electron-dense nucleus surrounded by a thin layer of cytoplasm is quite similar to the polar globules described in Actinomyxida (Marques, 1983). Its occurrence in *Paramyxa* may be related to meiosis.

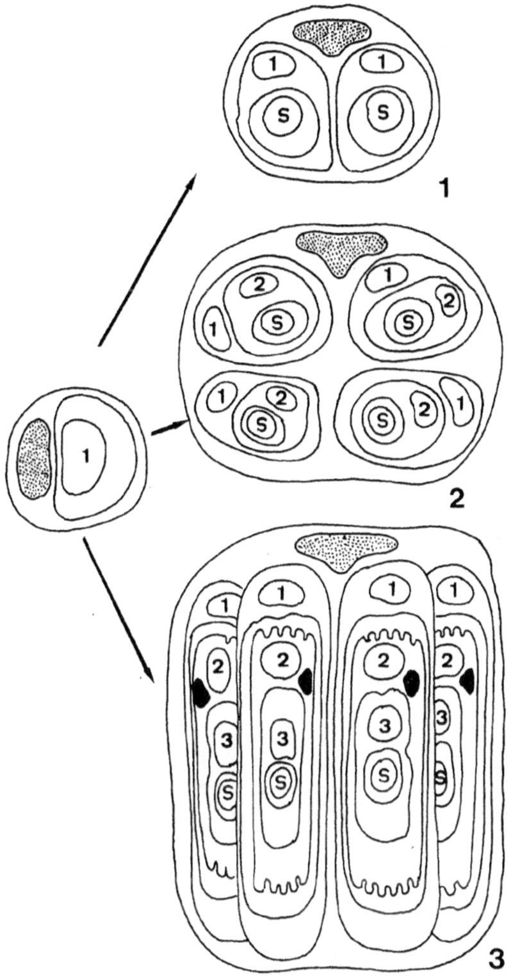

Fig. 7. The comparative evolution of the bicellular sporont (secondary cell + inner third cell) in the three paramyxean genera. The division of the third cell '1' produces 2 spores consisting of 2 cells in *Paramarteilia* (1), 4 spores made up of 3 cells in *Marteilia* (2) and 4 spores made up of 4 cells in *Paramyxa* (3). The dark element figured in secondary sporal cells of *Paramyxa* corresponds to the presumed polar globule. The dotted nucleus is that of the sporont. 1, 2, 3, S, nuclei of the primary, secondary, tertiary sporal cells and sporoplasm.

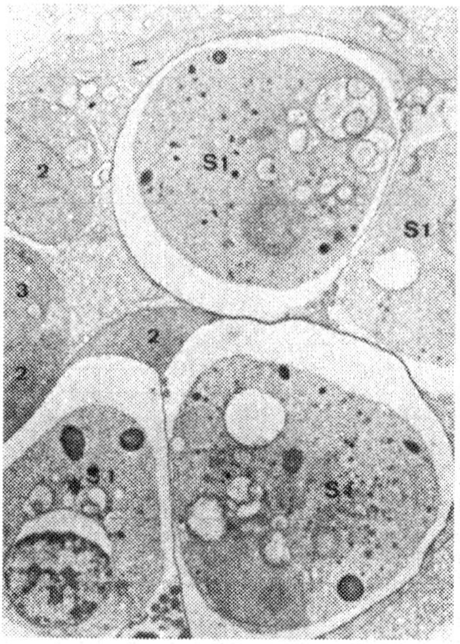

Fig. 8. The stem cell of *Paramarteilia orchestiae* with sporonts at different developmental stages. 2, nucleus of the sporonts (secondary cells); 3, nucleus of the tertiary cell; S1, primary sporal cells resulting from the division of the tertiary cell. × 6 410.

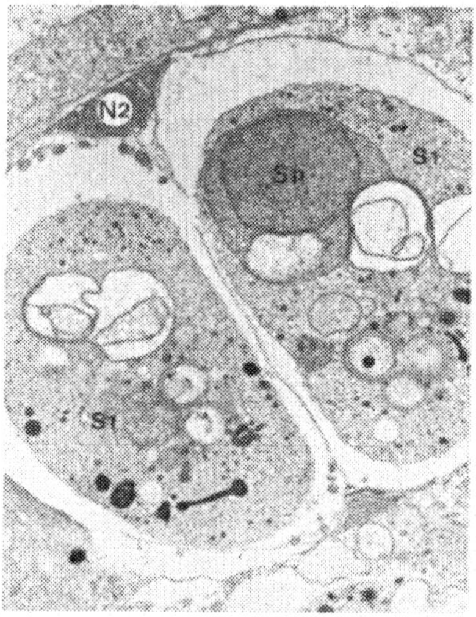

Fig. 9. A mature sporont of *Paramarteilia orchestiae*. The sporoplasm Sp is visible in one of the two primary sporal cells (S1); N2, sporont nucleus. × 16 000.

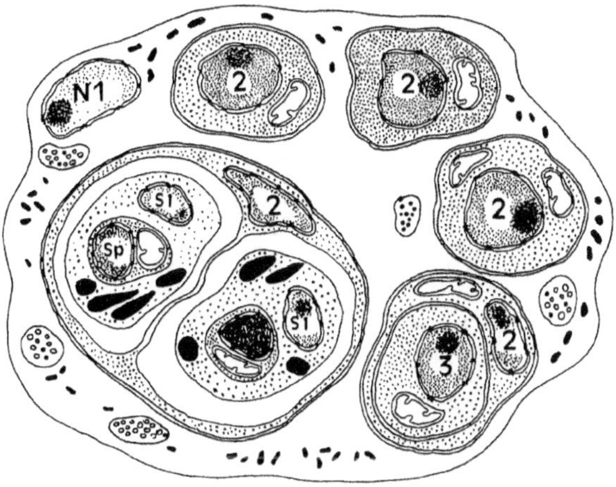

Fig. 10. The development of *Paramarteilia orchestiae*. N1, stem cell nucleus, 2, 3, nuclei of the secondary and tertiary cells; S1, primary sporal cell; Sp, sporoplasm.

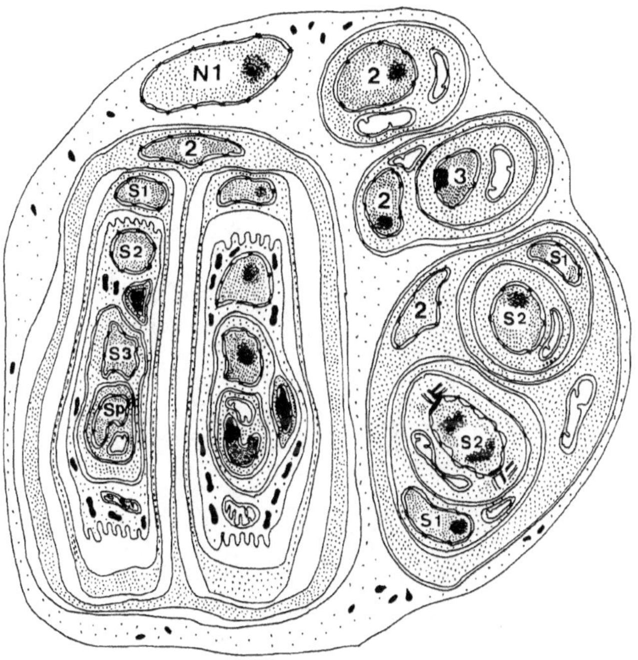

Fig. 11. The development of *Paramyxa paradoxa*. Only 2 of the 4 spores are figured in the young sporogenetic sporont (bottom right) and in the mature one (bottom left). S1, S2, S3, Sp, nuclei of sporal cells 1, 2, 3 and sporoplasm. The small dark cell visible in the mature spore is interpreted as the polar globule rejected before the differentiation of the haploid sporoplasm.

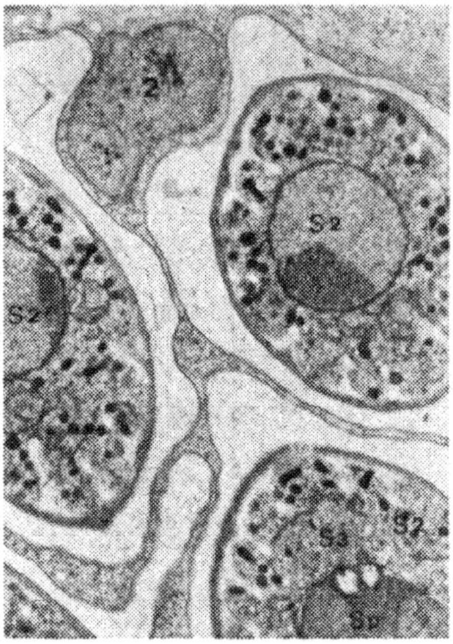

Fig. 12. A mature sporont of *Paramyxa paradoxa*. 2, sporont nucleus; S2, S3, secondary and tertiary sporal cells; Sp, sporoplasm. × 8 400 (after Desportes, 1981).

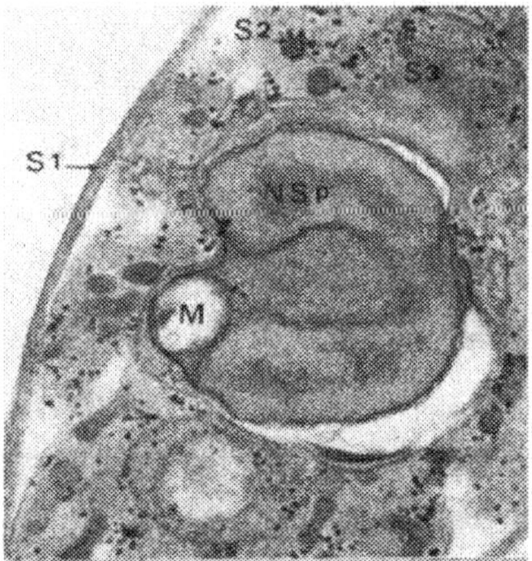

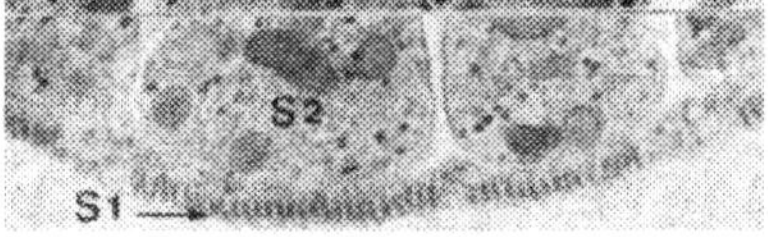

Fig. 13. The sporoplasm of *Paramyxa paradoxa* and its U-shaped nucleus. M, mitochondrion. S1, the thin cytoplasmic layer of the enveloping sporal cell (= primary sporal cell). Glycogen particles are contrasted by Thiéry reaction in the secondary and tertiary sporal cells (S2 and S3). × 21 600.

Fig. 14. The cytoskeleton of the primary sporal cell of *P. paradoxa* (S1); S2, secondary sporal cell. × 21 600.

3. Evolution Towards the Multicellularity and Cell Differentiation

The increased number of the sporal cells results in the concomitant acquisition of additional differentiating characters. In the highly differentiated mature spore of *Paramyxa*, the primary sporal cell, at first similar to the young secondary cell (dense cytoplasm and nucleus), is progressively reduced to a thin cytoplasmic layer mainly constituted by a tubular and fibrillar cytoskeleton (Figures 14 and 15). The secondary sporal cell, rich in dense bodies, is characterized by its peripheral infoldings of vesicular origin (Figures 12–15). Dense bodies, more or less elongated, also occur in the primary cell of *Paramarteilia* (Figures 8–10) and *Marteilia* (Perkins, 1976, Perkins and Wolf, 1976); elongated vesicles are present in the secondary sporal cell of the latter.

Only the sporoplasm is nearly identical in all genera. It is recognizable by its small size (1,5 μm – 2 μm), its cytoplasm rich in ribosomes, the dense nucleus and the occurrence of a single light-matrix mitochondria (Figures 9, 12, and 13). It is similar in ultrastructure to young secondary cells and other undifferentiated young cells. In *Paramyxa paradoxa* however, the sporoplasm nucleus is U-shaped (Figure 13).

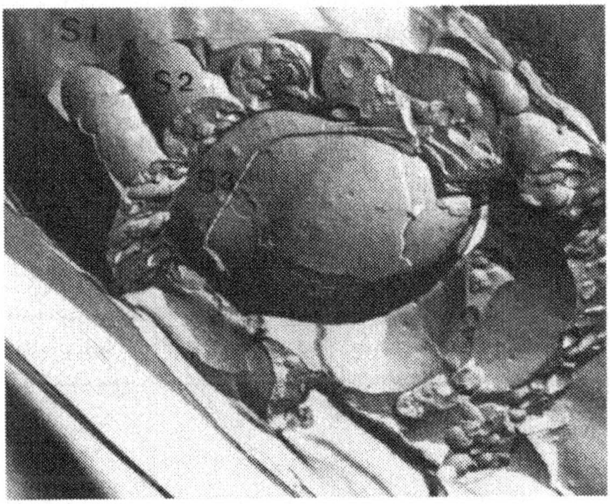

Fig. 15. Freeze-fracture of the spore of *Paramyxa paradoxa*. S1, the outer primary sporal cell; S2, the secondary sporal cell and its peripheral infoldings; S3, the tertiary sporal cell and possibly the inner sporoplasm. × 24 000 (after Desportes and Nicolas, research in progress).

4. Concluding Remarks

1. The acquisition of individual features occurs when the cells are enclosed within each other. The cell differentiation appears therefore as a consequence of the endogenous budding, a peculiar process which should be further studied.

2. The paramyxean development is homogenous in the whole group. *Paramyxa*

paradoxa seems to be the more differentiated parasite, considering the higher number of its sporal cells and their higher individualization.

The following classification is proposed:

The PARAMYXEA Levine. (This class will be probably raised to the rank of a phylum)
Characterizations: Sporulation resulting from a series of endogenous buddings within a stem cell. Spores consisting of several cells enclosed inside each other. Nine singlet microtubules in centrioles. All members parasitic in marine invertebrates. Two orders:

Order 1. PARAMYXIDA, Chatton (1911)
– possible occurrence of a meiosis with elimination of a polar globule.
Paramyxa paradoxa parasitic in Polychaeta larvae (*Poecilochaetus serpens*).
Order 2. MARTEILIIDA, Desportes and Ginsburger-Vogel (1977)
– not any indication of a meiosis.
Two genera: 1. *Marteilia* (3 species) parasitic in oysters
2. *Paramarteilia* (one species) parasitic in Crustacea

3. The Paramyxea exhibit both protistan and metazoan characters. Some features are protistan ones: the occurrence of the whole development within a stem cell and the structure of the centriole. The latter is made of nine singlets instead of the classical triplets (Figure 4). This centriolar structure corresponds to the procentriole (or prokinetosome) described in ciliates and flagellates as well as in sporozoa where the triplet organization has never been observed (see Desportes and Ginsburger-Vogel, 1977).

However the successive production of degenerative cells, except the sporoplasm, as well as the cell differentiation observed in the multicellular spores are metazoan characteristics.

The main novelty of Paramyxea is that the cell differentiation results from the enclosing of the cells inside each other instead of being associated side by side as in Metazoa. It is why the Paramyxea may be somewhat considered as 'ablastic metazoa'.

References

Committee on Systematics, 1980, *J. Protozool.* **27**, 37.
Desportes, I.: 1981, *Protistol.* **17**, 365.
Desportes, I. and Ginsburger-Vogel, T.: 1977, *Protistol.* **13**, 607.
Desportes, I. and Lom, J.: 1980, *C. R. Acad. Sci. Paris* **292**, 627.
Ginsburger-Vogel, T. and Desportes, I.: 1979, *J. Protozool.* **26**, 390.
Ginsburger-Vogel, T., Desportes, I., and Zerbib, Ch.; 1976, *C. R. Acad. Sci. Paris* **283**, 939.
Grizel, H., Comps, M., Bonami, J. R., Cousserans, F., Duthoit, J. L., Le Pennec, A. M.: 1974, *Sci. Pêche* **240**, 7.
Marquès, A.: 1983, *C. R. Acad. Sci. Paris* **296**, 717.
Perkins, F. O.: 1976, *J. Protozool.* **23**, 64.
Perkins, F. O. and Wolf, P. H.: 1976, *J. Parasit.* **62**, 528.

TABLE OF CONTENTS

ARTICLES

TABLE OF CONTENTS

TABLE OF CONTENTS